Metamaterials - History, Current State, Applications, and Perspectives

Edited by Aleksey Kuznetsov

Published in London, United Kingdom

Metamaterials - History, Current State, Applications, and Perspectives
http://dx.doi.org/10.5772/intechopen.102177
Edited by Aleksey Kuznetsov

Contributors
Jiangyi Zhang, Costantino De Angelis, Davide Rocco, Andrea Locatelli, Domenico De Ceglia, Andrea Tognazzi, Attilio Zilli, Michele Celebrano, Marco Finazzi, Antonio Ferraro, Roberto Caputo, Andrews Christina Josephine Malathi, Mohamed Lashab, Mounir Belattar, Sekkache Hocine, Halfaya Ahmed, Alkesh Agrawal, Babar Kamal, Usman Ali, Jingdong Chen, Sadiq Ullah

First published in London, United Kingdom, 2023 by IntechOpen
IntechOpen is the global imprint of INTECHOPEN LIMITED, registered in England and Wales, registration number: 11086078, 5 Princes Gate Court, London, SW7 2QJ, United Kingdom

British Library Cataloguing-in-Publication Data
A catalogue record for this book is available from the British Library

Additional hard and PDF copies can be obtained from orders@intechopen.com

Metamaterials - History, Current State, Applications, and Perspectives
Edited by Aleksey Kuznetsov
p. cm.
Print ISBN 978-1-80356-809-6
Online ISBN 978-1-80356-810-2
eBook (PDF) ISBN 978-1-80356-811-9

Meet the editor

Dr. Aleksey Kuznetsov obtained his Ph.D. in Physical Chemistry from the Department of Chemistry and Biochemistry, Utah State University, USA. He graduated in 2003, after 3 years of doctorate studies, with a specialization in computational/theoretical chemistry. In 2019, after several postdoctoral and visiting professor positions in Germany (including an Alexander von Humboldt fellowship stay), the United States, and Brazil, he acquired a permanent faculty position at the Department of Chemistry, Universidad Técnica Federico Santa Maria, Chile, where he has been working on the computational design of various complexes of porphyrins, including core-modified porphyrins, with nanoparticles, fullerenes, and graphenes, along with computational studies of transition metal complexes, organic compounds with pharmacological applications, metal–fullerene complexes, compounds with anticorrosive properties, and more.

Contents

Preface

Metamaterials are a novel family of materials that has been attracting increased interest from researchers during the last decades. Metamaterials are engineered in such a way that they possess various properties of interest not found in naturally occurring materials [1]. The name "metamaterials" originates from two words: the Greek word μετά (*meta*), meaning "beyond" or "after," and the Latin word *materia*, meaning "matter" or "material" [1]. These materials are built of specially designed assemblies of multiple elements made from various composites, which can contain metals and plastics. These assemblies, in turn, are arranged in repeating patterns designed in such a way that they have size scales smaller than the wavelengths of the phenomena they should influence. Thus, metamaterials have their novel and unusual properties derived not from their base materials' properties or combinations of those but rather from their newly designed structures.

The shapes, geometries, sizes, orientations, and arrangements of metamaterials give rise to their "smart" properties, making the devices built of metamaterials capable of selective and careful manipulation of electromagnetic waves of various wavelengths. This can be achieved by absorbing, enhancing, blocking, or bending waves, resulting in metamaterials having benefits well beyond the characteristics of conventional materials.

Applications of metamaterials are ever growing and diverse. They can be used in optical filters, medical devices, remote aerospace engineering applications, security areas such as sensor detection, infrastructure monitoring, and crowd control, smart solar power management, lasers [2], radomes (structural weatherproof enclosures protecting radar antennae), high-frequency battlefield communication, lenses for high-gain antennas, to improve characteristics of ultrasonic sensors, and even in creating shielding structures for protection from earthquakes [3–6]. Furthermore, super-lenses, which can allow obtaining images below the diffraction limit that can be achieved by conventional lenses, can be manufactured from metamaterials [7]. Also, sub-wavelength optical metamaterials integrated with optical recording media can allow achieving optical data density higher than that limited by diffraction [8]. Moreover, metamaterials can render humans or objects a form of "invisibility" and can be applied to acoustic and seismicity devices [3, 9].

Thus, metamaterials research is an extremely interesting and promising interdisciplinary area of science and technology that involves various fields of knowledge such as electrical engineering, electromagnetics, classical optics, solid-state physics, microwave and antenna engineering, optoelectronics, material sciences, nanoscience, and semiconductor engineering [10]. This book contributes to broadening and enriching our knowledge and understanding of these materials, some specific aspects of their state of the art, and further perspectives.

Chapter 1, "Applications of Metamaterials and Metasurfaces" by Babar Kamal et al., presents a brief overview of the known types and applications of metamaterials/metasurfaces followed by a comprehensive analysis of these surfaces for antenna performance enhancement, polarization conversion, radar cross-section reduction, and wave absorption.

Chapter 2, "Study on Miniaturization of Antenna Using Metamaterials" by Andrews Christina Josephine Malathi, discusses applications of metamaterials, focusing primarily on their antenna application.

Chapter 3, "Manipulating Light with Tunable Nanoantennas and Metasurfaces" by Costantino De Angelis et al., reviews recent achievements in the field of tunable metasurfaces and presents a platform based on an AlGaAs metasurface that allows the modulation of the generated second harmonic signal.

Chapter 4 "Metamaterial Applications in Modern Antennas" by Mohamed Lashab et al., presents different types of metamaterials, their historical evolution, physical properties, and applications in antenna design.

Chapter 5 "An Application-Based Study on Electromagnetic Absorber Using Metamaterial" by Alkesh Agrawal, focuses on the classification of materials on the basis of permittivity and permeability and on the specific study of the electromagnetic absorber.

Finally, Chapter 6 "Amplitude-Dependent Acoustic Absorber" by Jiangyi Zhang, considers the design of a 1D amplitude-dependent acoustic absorber.

Aleksey Kuznetsov
Department of Chemistry,
Universidad Técnica Federico Santa Maria,
Santiago, Chile

References

[1] Kshetrimayum RS. A brief intro to metamaterials. IEEE Potentials. 2004;**23**(5):44-46. DOI: 10.1109/MP.2005.1368916. S2CID 36925376

[2] Awad E. A novel metamaterial gain-waveguide nanolaser. Optics & Laser Technology. 2021;**142**:107202

[3] Brun M, Guenneau S, Movchan AB. Achieving control of in-plane elastic waves. Applied Physics Letters. 2009;**94**(61903):061903. DOI: 10.1063/1.3068491. S2CID 17568906

[4] Rainsford, Tamath J, Abbott D, Derek A. T-ray sensing applications: Review of global developments. In: Al-Sarawi Said F, editor. Proc. SPIE. Smart Structures, Devices, and Systems II. 5649 Smart Structures, Devices, and Systems II (Poster Session). 2005. pp. 826-838. DOI: 10.1117/12.607746 S2CID 14374107

[5] Cotton MG. Applied Electromagnetics. Technical Progress Report (NITA – ITS). Telecommunications Theory (3). 2003. pp. 4-5

[6] Boratay AK, Ekmel Ö. Radiation properties of a split ring resonator and monopole composite. Physica Status Solidi B. 2007;**244**(4):1192-1196

[7] Guerra JM. Super-resolution through illumination by diffraction-born evanescent waves. Applied Physics Letters. 1995;**66**(26):3555-3557

[8] John G, Dmitri V, Paul S, Walter H, Lukas T. Near-field optical recording without low-flying heads: Integral near-field optical (INFO) media. Japanese Journal of Applied Physics. 2002;**41**(Part 1, 3B):1866-1875

[9] Guenneau SB, Movchan A, Pétursson G, Anantha Ramakrishna S. Acoustic metamaterials for sound focusing and confinement. New Journal of Physics. 2007;**9**(11):399

[10] Saïd Z, Sihvola A, Vinogradov AP. Metamaterials and Plasmonics: Fundamentals, Modelling, Applications. New York: Springer-Verlag; 2008. pp. 3-10, Chap. 3, 106 ISBN 978-1-4020-9406-4

Section 1

Metamaterials: General Applications

Chapter 1

Applications of Metamaterials and Metasurfaces

Babar Kamal, Usman Ali, Jingdong Chen and Sadiq Ullah

Abstract

Metamaterials are efficiently homogenizable arrangements of artificial structural components engineered to achieve beneficial and exotic electromagnetic (EM) properties not found in natural materials. Metasurfaces are the two-dimensional analogue of metamaterials consisting of single-layer or multi-layer stacks of planar structures. Both metamaterials and metasurfaces have great potential to be used in a wide range of applications, e.g., antennas, polarization converters, radar cross section (RCS) reduction, and absorbers, to control the amplitude, phase and polarization of the reflected and transmitted EM waves. This chapter presents a brief overview of the known types and applications of metamaterials/metasurface followed by comprehensive analysis of these surfaces for antennas performance enhancement, polarization conversion, RCS reduction, and wave absorption.

Keywords: metasurface, electromagnetic waves, antennas, polarization, absorption

1. Introduction

Metamaterials are artificial three dimensional (3D) structures composed of periodic subwavelength metal/dielectric arrangement of unit cells with exact dimensions [1]. A metamaterial is volumetric and is intended to provide artificial permeability (μ) and permittivity (ε) with extremely powerful electromagnetic (EM) wave control capabilities [2], leading to exceptional EM properties, which are often provided via resonant effects managed by the geometry of the unit cells [3]. Metamaterials have been used in such applications as magnetic resonance imaging [4], sensors [5], and antenna systems (for gain enhancement and beam steering) [6]. For EM integration, bulky sizes, low efficiencies, complex fabrication process are some of the disadvantages of metamaterials. To overcome these disadvantages, the two-dimensional (2D) metamaterials, which are known as metasurfaces, have been developed. They have attracted increasing research and development over recent years due to their small size, larger bandwidth, high efficiency, ease of fabrication, and less lossy characteristics [7–9]. A metasurface (MS) is intended to manipulate the amplitude, phase, and polarization of incident EM waves. Traditional techniques achieve polarization control of EM waves through using the crystal optical activity and Faraday effect. The drawback associated with these techniques are narrow bandwidth and incident angle response, and the resulting devices are large in size [10]. Similarly, the conventional absorbers, which are used to achieve absorption of EM waves, are bulky, fragile and challenging to manufacture. Metasurfaces can be utilized to control the EM waves within one thin layer, leading to substantial advantages, such as low cost, and high degree of integration. They can be used in a wide spectrum of applications such as perfect lens [11], perfect absorbing [12], invisibility [13], light bending [14], and polarization conversion [15], to name but a few.

In this chapter, various metamaterials are discussed for antenna systems to achieve high gain as well as wide bandwidth, size reduction and reduced Specific Absorption Rate (SAR). Also discussed are metasurfaces with unique geometrical arrangements to achieve polarization conversion, RCS reduction, and absorption of EM waves. The basic design requirements, fundamental principle, working mechanism, parametric extraction and unit cell configurations are discussed in detail to explain the core concept of metamaterials/metasurfaces.

2. Applications of metamaterials

2.1 Metamaterials in the design of antenna

The metamaterials have unique physical characteristics that are not found in nature but are highly desirable for use in many applications including but not limited to invisible submarines, microwave invisibility cloaks, compact effective antennas, and negative refractive-index lenses [16]. One of the most important applications of metamaterials is in antenna design [17]. Due to the distinctive characteristics of metamaterials, novel antennas can be designed that are not possible with the materials found in nature. The metamaterial integrated in antennas, which consist of a single layer or double layers, can either serve as substrates or be added to the geometry of the antenna to improve the overall performance of the antenna [18, 19]. Scientific investigations showed that introducing metamaterials into the antenna design can improve several important parameters and decrease the overall volume of the antenna while increasing the radiated power and improving gain and directivity [20, 21]. Moreover, these materials can be integrated into antennas to reduce the side and back lobe radiation as well as to restrict the specific absorption rate (SAR) in scenarios where they are worn on the body [22, 23]. Depending on how the antenna is designed, different construction and applications of these materials can be used.

2.1.1 Metamaterial unit cell

The metamaterials used in antennas can be of a single unit cell or an array of unit cells. Therefore, the most crucial step in creating the metamaterials is the design of the unit cell. Its key features that affect the permittivity (ε), permeability (μ) and resonance frequency (f_r) must be considered and analyzed [24], based on which as well as the requirements at the desired frequencies, various parameters, e.g., the size and shape of the unit cell, its ε, μ, and resonant frequencies are obtained. To satisfy the given requirements at the resonant frequency, the dimensions and type of unit cell are optimized [25]. The unit cell size may vary greatly depending on the geometry of the metamaterials, but it is generally less than one-tenth of the working wavelength [26], as depicted in **Figure 1**.

For instance, several metamaterial unit cells at the 2.4 GHz Industrial Scientific and Medicinal (ISM) frequency band are designed and simulated using the procedures

DOI: http://dx.doi.org/10.5772/intechopen.108145

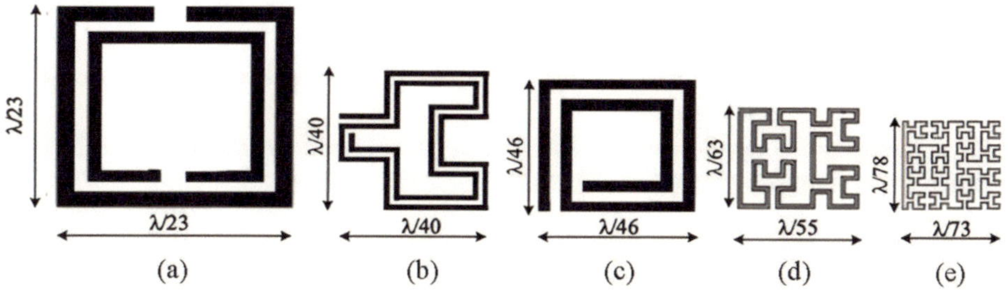

Figure 1.
Layout of a unit cell of an addition with square ring resonator (SRR) (a), the addition of a 2nd order Hilbert fractal (b), a square spiral (c), the addition of a 3rd order Hilbert fractal (d), the addition of a 4th order Hilbert fractal inclusion (e) with the increase of order the size is reduced [26].

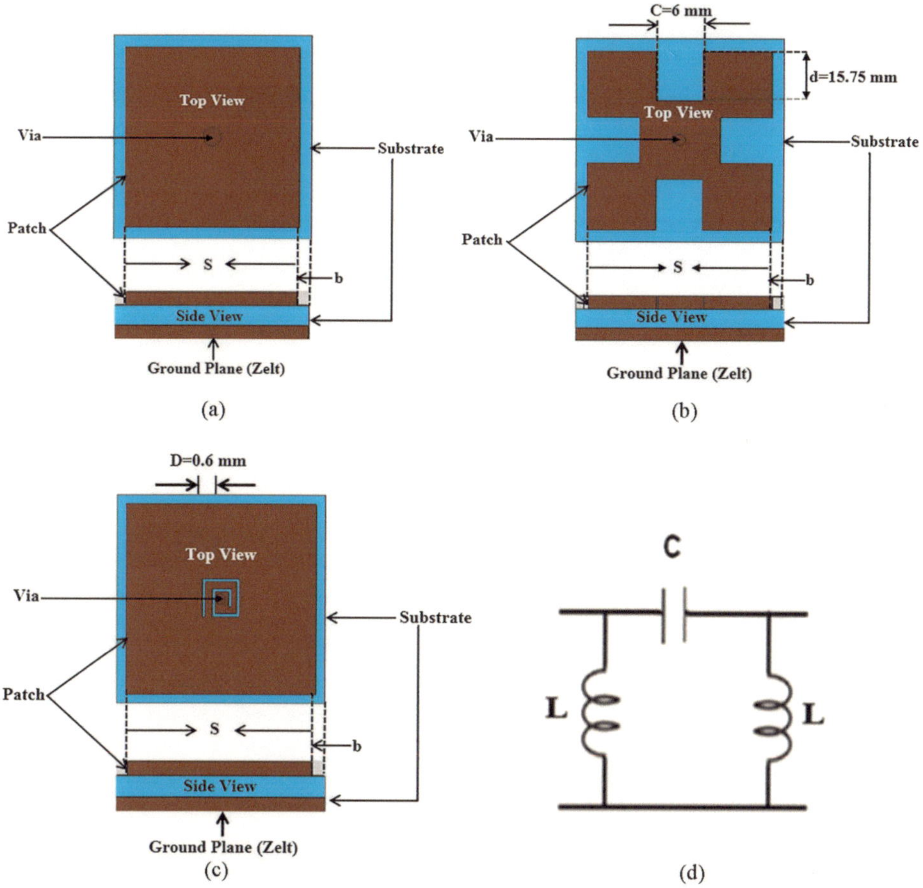

Figure 2.
Geometry of: (a) the mushroom-like structure, (b) the Slotted structure, (c) the Spiral structure. (d) The LC-resonant circuit of the unit cell [23].

listed below. To meet the homogeneous requirements by metamaterials, it is critical that the size of the unit cell needs to be smaller than the guided wavelength (**Figure 2**).

Figure 2 illustrates the geometry of the different unit cells. These structures function as an LC parallel resonant circuit. The capacitance C and inductance L, whose value depends on the unit cell's geometry and size [27], determine the in-phase bandwidth and resonant frequency.

The capacitance is due to the fringing between adjacent unit cells. It can be calculated as:

$$C = \frac{w\varepsilon_{0(1+\varepsilon_r)}}{\pi} \cosh^{-1} \frac{(w+g)}{g} \tag{1}$$

where w, g and ε_0 represent the width of the unit cell, the gap between the neighboring unit cells, and the dielectric constant of free space, respectively. The following relationship can be used to calculate the inductance L, which is directly related to the substrate's thickness or the length of the metal via.

$$L = \mu_0 \mu_r t \tag{2}$$

The optimized unit cells dimensions at the desired 2.4 GHz frequency is shown in **Figure 2**. The dimensions are: $w = 0.323\lambda_{2.4\ \text{GHz}}$, $g = 0.018\lambda_{2.4\ \text{GHz}}$, where $\lambda_{2.4\ \text{GHz}}$ is the free space wavelength at 2.4 GHz, via (radius) = $0.004\lambda_{2.4\ \text{GHz}}$. The mentioned surfaces are categorized for surface wave suppression and in-phase reflection characteristics. The unit cell's resonance frequency is then given as:

$$f_r = \frac{1}{2\pi\sqrt{LC}} \tag{3}$$

2.1.2 In-phase reflection

The simulation setup and boundary conditions for in-phase reflection analysis of the three metamaterial structures are shown in **Figure 3**. Linearly polarized (TE10) plane waves are used to excite these surfaces. The in-phase reflection (0°) at the resonant frequency of 2.4 GHz is determined by the three unit cells (**Figure 4**). The

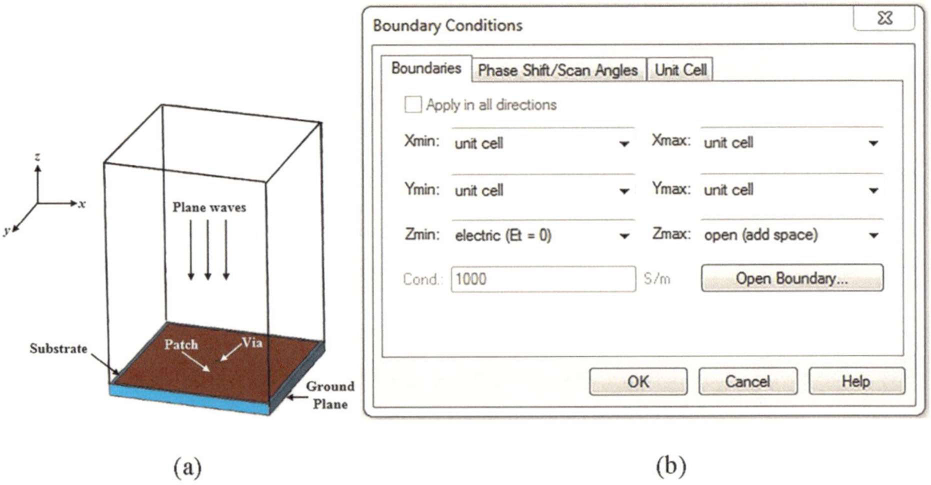

Figure 3.
Simulation setup of the unit cell for the in-phase reflection analysis [23].

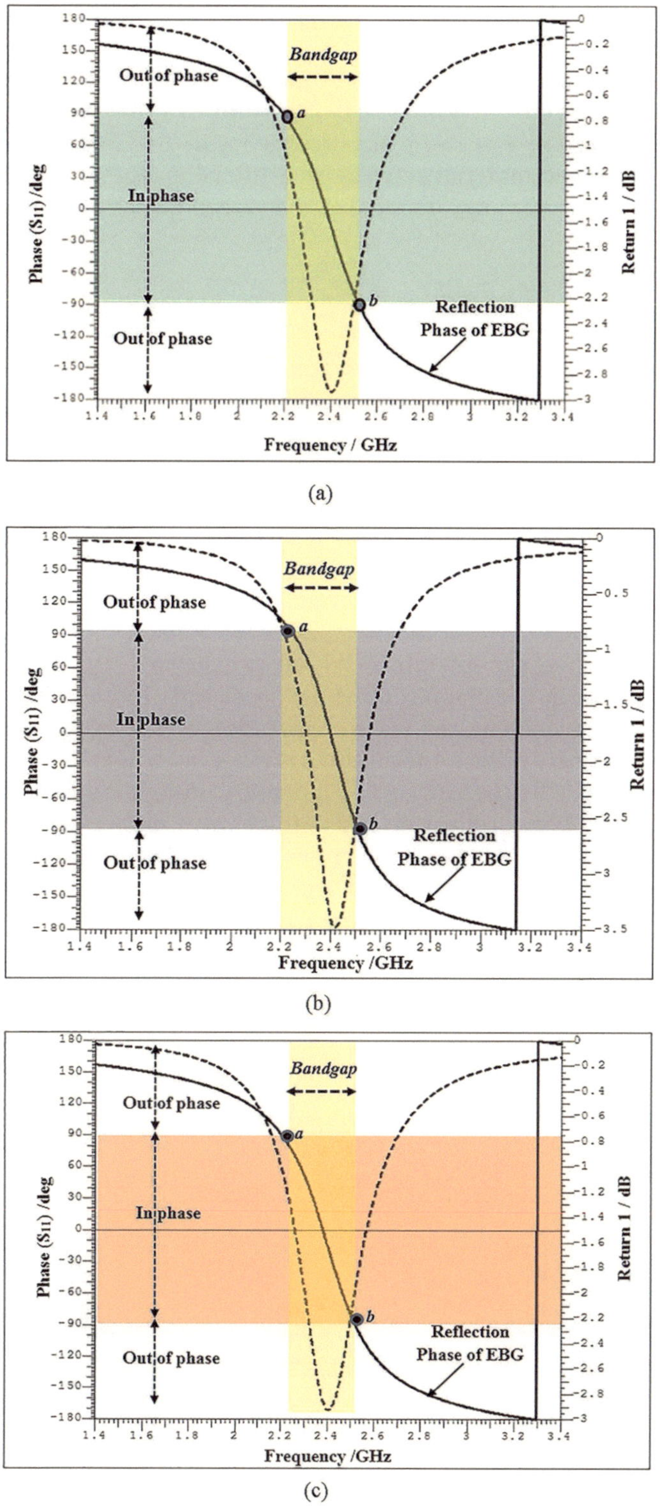

Figure 4.
Reflection phase characteristics of three structures: (a) Mushroom-like, (b) Slotted, and (c) Spiral [23].

surface acts like a perfect magnetic conductor (PMC) at this frequency [23]. At a lower frequency of 1.4 GHz and a higher frequency of 3.4 GHz, the reflection phase is +180° and −180°. At these two frequencies, the surface acts like a perfect electric conductor (PEC). The reflection phase changes from +90° to −90° at points "a" and "b", respectively, crossing 0 at the central frequency of 2.4 GHz. In the given reflection bandwidth, the metamaterial acts like an artificial magnetic conductor (AMC). The reflection phase bandwidth, denoted as BW_{rp}, can be calculated as follows:

$$BW_{rp} = \frac{f_b - f_a}{f_c} \times 100, \quad (4)$$

where f_b, f_a, and f_c denote, respectively, the higher, lower, and centeral frequencies.

2.1.3 Surface wave bandgap

To analyze the surface wave bandgap characteristics, a 50-Ω suspended microstrip line at a distance of 1.5 millimeters above the array of unit cells is used (**Figure 5**). To examine the transmission coefficient of the metasurface, the microstrip line is powered using a waveguide port at both ends. Note that the dispersion diagram method can also be used to study the bandgap characteristics.

The S_{12} and S_{11} of the aforementioned three metasurface structures are plotted in **Figure 6**. It is found that the magnitude of the S_{12} directly influences the surface wave suppression. If setting the required wave suppression threshold of S_{12} to −20 dB, the mushroom-like structure efficiently suppresses the surface waves in the frequency band from 2.34 to 2.57 GHz according to **Figure 6**. Consequently, this type of structure provides both in-phase reflection and surface wave suppression characteristics at the resonant frequency. This surface functions as a standard PEC ground plane outside of the 2.34–2.57 GHz frequency band. According to the pre-defined criteria in this band, the remaining two metamaterial structures, i.e., the spiral and the slotted, exhibit in-phase reflection characteristics only at 2.4 GHz but without sufficient surface wave suppression. The surface wave bandgap of the spiral metasurface shifts slightly to the left of the 2.4 GHz point as a result of the inductive loading of this surface. CST Microwave Studio software [23] is used to study and calculate the unit cell's properties, which show that the unit cell clearly serves as an AMC.

Similarly, a dual band metasurface unit cell is considered and examined using the following steps. The layered structure is shown in **Figure 7**. The dual-band unit cell consists of three layers. A square dielectric block has been placed between the first and

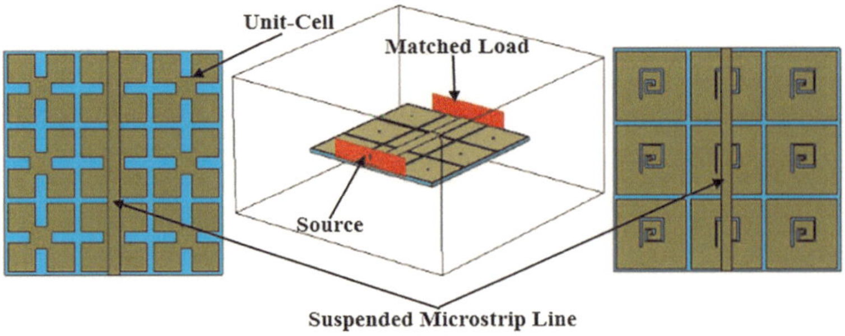

Figure 5.
Simulation setup for the surface wave bandgap characterization [23].

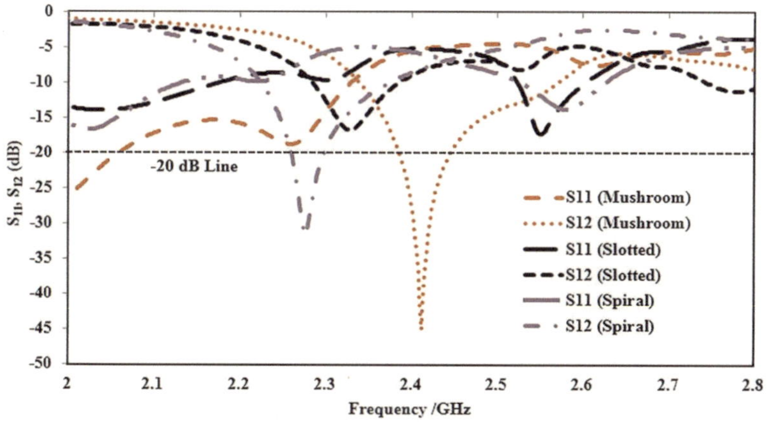

Figure 6.
S_{11} and S_{12} characteristics of three metamaterial structures [23].

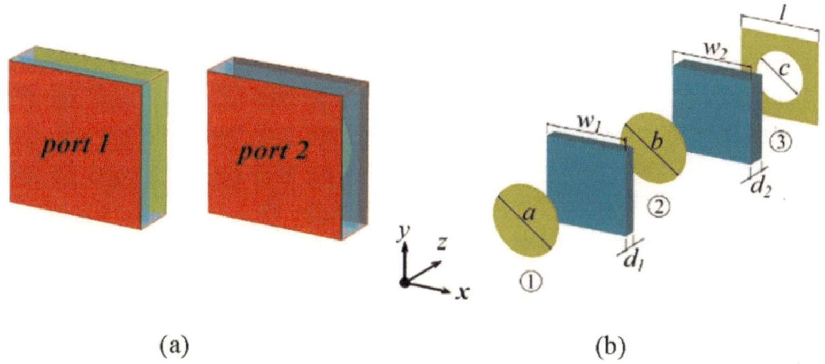

Figure 7.
(a) Layout of the Unit Cell from ports 1 to 2. (b) Hierarchical diagram of EM incidence waves [28].

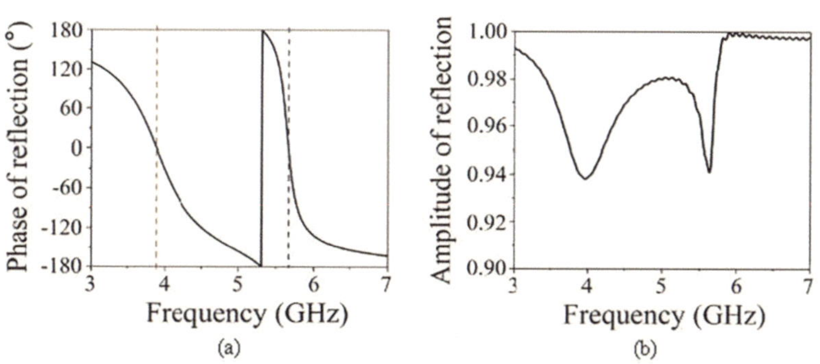

Figure 8.
Simulated results for a unit cell: (a) phase of the reflection coefficient, and (b) amplitude of the reflection coefficient [28].

second layers, which are made up of circular metal patches with diameters of a and b, respectively. The third layer consists of a square metal patch with a central circular slot. Between the second and third layers, a square dielectric block is inserted [28].

The magnetic and electric wall boundary conditions were forced along the $\pm x$ and $\pm y$ directions to produce a waveguide. Then, the circular metal patch's surface received port 1 of the waveguide and its reference plane, and the square metal patch's surface received port 2 of the waveguide with the circular slot. Due to the linear polarization along the y-direction, excitation of port 1 is comparable to a plane wave incident along the z-direction. **Figure 8(a)** shows the unit cell during its reflection phase. Phase is zero at 3.89 and 5.66 GHz. At these two frequencies, the unit cell is equivalent to mu-negative metamaterials (MNMs). **Figure 8(b)** shows that at these two operating frequencies, the reflection coefficient is approximately 0.94, which demonstrates the partial reflection property of the unit cell.

3. Metamaterials in antennas engineering

In this section, the application of metamaterials for enhancing antenna parameters is presented and explained. Antennas play an important role on wireless communication while materials play a crucial role on the antenna performance. The widely used microwave and radio frequency substrate materials in antenna applications include artificial magnetic conductors (AMCs) and high-impedance surfaces (HISs). HISs or AMCs are used to build small, low-profile antenna systems by placing them close to or all around the antenna radiating elements. Metamaterials can also be used to build the antenna or as part of the feeding mechanism for the antenna system.

What discussed in this section includes gain and directivity enhancement, bandwidth enhancement, surface wave suppression and SAR reduction of the antenna when placed in close proximity to the human body, and miniaturization of microstrip antennas by loading the materials as a patch, mounting them as a superstrate above the main radiator, placed them at the ground plane, or embedding them in the substrate. These surfaces are used in antennas to solve a variety of problems, thereby circumventing the limits of the antennas designed with the conventional approaches. For example, placing the material above the reflector can help enhance the radiation properties of the antenna, reduce the backward radiation and redirect the radiation in the forward direction [29]. The electromagnetic band gap (EBG) structure, first proposed by Sievenpiper in 1999 [30], was a mushroom-shaped structure with a ground plane loaded with a grid of square patches attached to the ground plane by a metallic through (**Figure 9**).

3.1 Incorporating metamaterial into antenna design

By incorporating metamaterial structures into the design of antennas, it is possible to decrease the antenna size, increase its gain and directivity, expand the bandwidth, and reduce side and back radiation. Metamaterial structures can also be used to create multiband applications. This section focuses on explaining the benefits of incorporating metamaterial structures into the antenna design.

3.1.1 Gain enhancement

The performance degradation in gain, directivity, and efficiency is the fundamental disadvantage of small planar antennas. In the case of multiband applications, this issue becomes more severe, particularly in the lower frequency bands. To meet the requirements for the total link budget of the transceiver systems, antennas must

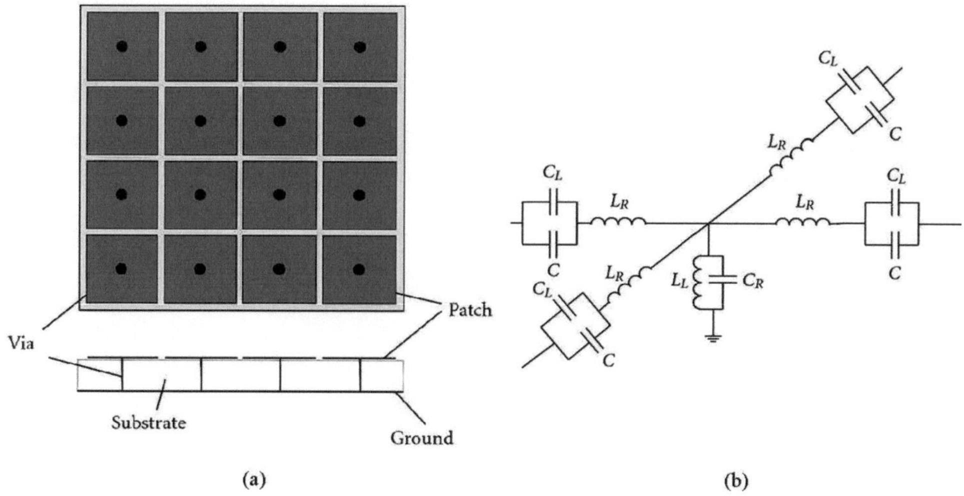

Figure 9.
(a) Layout of the mushroom electromagnetic band gap (EBG) unit cell, (b) equivalent circuit model of the unit cell [31].

overcome the problem of low gain and efficiency. Besides using array antennas, metasurface antennas were recently introduced as an alternative candidate for various communication bands to improve the overall performance of the antenna by incorporating metamaterial structures (either AMM or AMC) into antenna designs.

Metamaterial are integrated into the antenna by either arranging their unit cells in such a way that they surround the antenna radiating element [23, 32], or using them as a ground plane loading or etching of the antenna, in which AMC functions as a zero-degree reflection for the incident waves at the antenna working frequency [33–36]. There are different ways to use metamaterial structures for enhancing the antenna gain, as illustrated in **Figure 10**. However, regardless of what method is used, the unit cell type, the number of superstrates used, and the distance between the primary radiating elements and the superstrate all play an important role on the gain improvement. The unit cells can be arranged in different ways, e.g., surrounded by the radiating elements of the antenna or loaded on both or one side of the substrate. In order for the metamaterial to have unique physical properties that fit the resonance frequency of the antenna, the size of the unit cells needs to be properly designed. Due to their negative permeability characteristics, the unit cells can be simply integrated with radiating components and used as insulators to reflect surface waves. The traditional antenna's gain can be improved by adding metamaterial unit cells around it [37]. The two key determinants that affect the gain attained are the resonant frequency of the desired antenna and the number of unit cells used. The array of unit cells with various relative permittivity is loaded away from the primary radiating element in the case of metamaterial superstrates. These unit cells can be loaded on either side of the superstrate. The number of superstrates, the number of unit cells, and the distance between the radiating element and the superstrate all play a key role on the gain performance and directivity of the conventional antenna [38] (**Figure 11**). The directivity and gain of the traditional antenna can be significantly increases by integrating metamaterial into the antenna design. However, the overall size and thickness of the antenna are also increased.

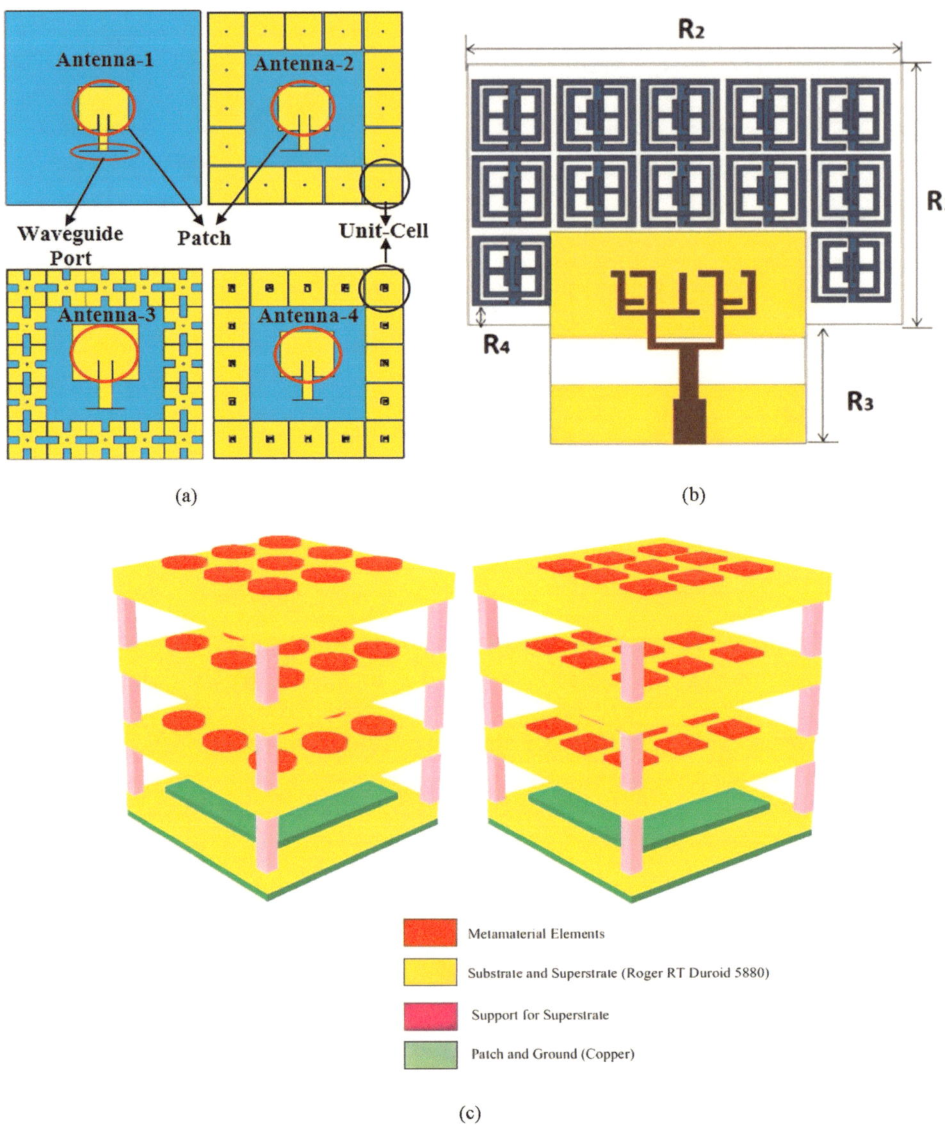

Figure 10.
Application of metamaterials to improve the antenna gain, including: (a) radiating element surrounded by unit cells, (b) antenna loading with metamaterials, and (c) metamaterial as a superstrate [23, 33, 34].

3.1.2 Metamaterials for bandwidth enhancement

In addition to improving the gain and directivity of conventional antennas, metamaterials can also be used to increase the impedance bandwidth of the antennas. To increase the bandwidth, the metasurfaces can be utilized as part of the antenna or can be used as a superstrate placing over the main radiating element, as was discussed for increasing the gain of the antenna (**Figure 11**). The superstrate can be placed above or below the metamaterial unit cells. The number of unit cells used and the distance between the radiating element and the superstrate have a great impact on the antenna impedance bandwidth. The configuration of the antenna with superstrate material and

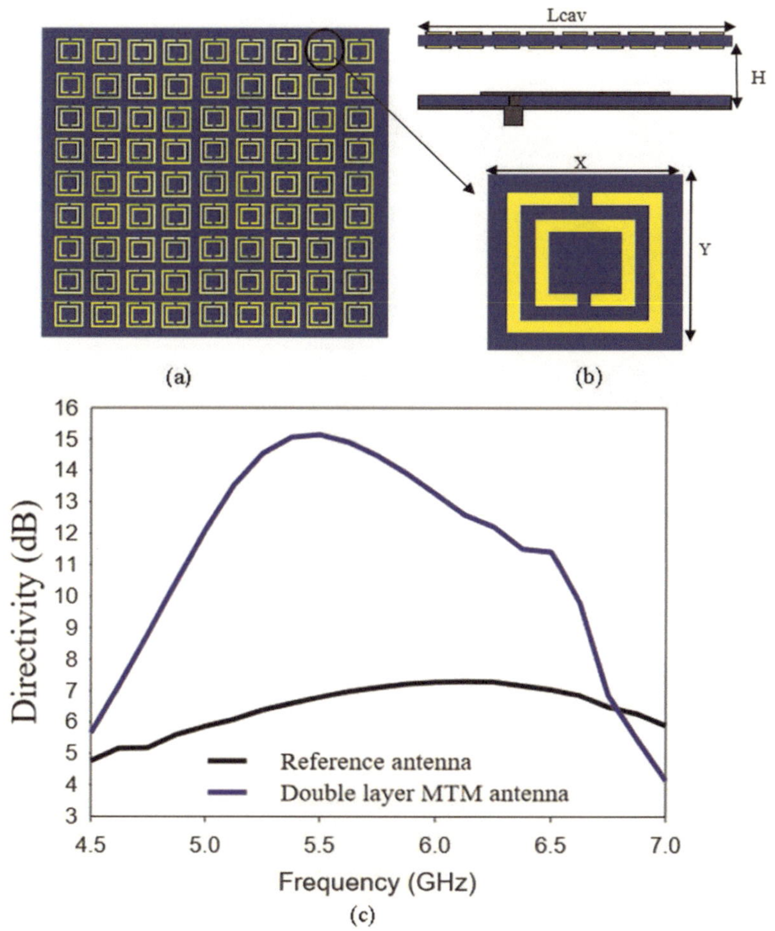

Figure 11.
Patch antenna with a double layer metamaterial superstrate: (a) front and top view of the antenna, (b) unit cell of SRR, and (c) directivity with/without superstrate [38].

the simulation of the reflection coefficient with and without the metamaterial are shown in (**Figure 12**). The bandwidth for the higher frequency band is significantly increased by the antenna integrated with metamaterial, from only 9.35–28%. Moreover, the third band has increased to 27.81% and moved slightly higher in frequency. Incorporating the metamaterial unit cells also matches the lower band [39].

3.1.3 Metamaterials for antenna size reduction

To reduce the antenna size, numerous design strategies are proposed, including shorting pins, generating disturbances, employing high dielectric substrate materials, fracturing the shape, etc. Many antenna designers have recently utilized metamaterial structures as defective ground structures (DGS) for small antennas. Using these structures as a DGS can produce unusual characteristics at the resonance frequency of the antenna. In this instance, the unit cell size is the same as the size of the DGS's removed portion [40, 41]. **Figure 13** depicts the geometry of the antenna and the metamaterial loaded with the bottom layer. The simulated S_{11} of the antenna with and

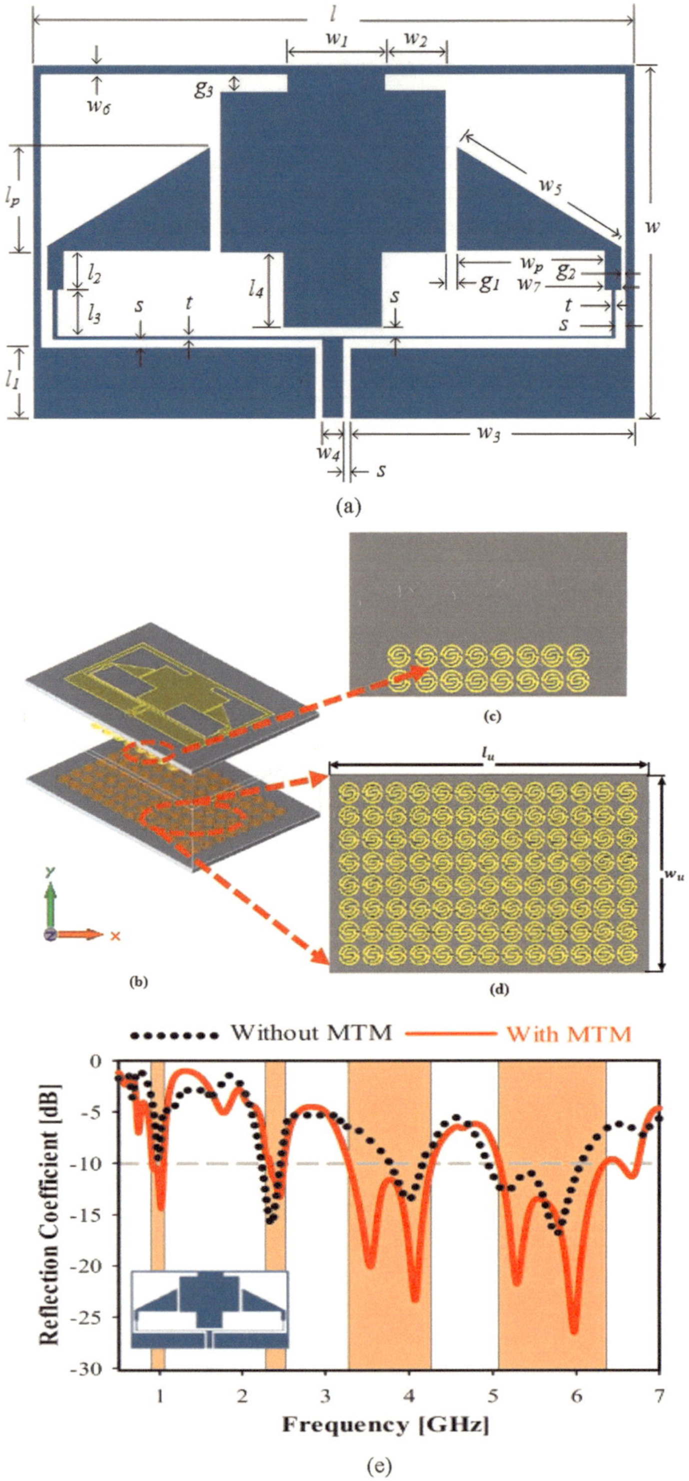

Figure 12.
Layout of the antenna (a) front view, (b) 3D view, (c) back view, (d) a suspended separator MMT layer, (e) S_{11} *with and without metamaterial [39].*

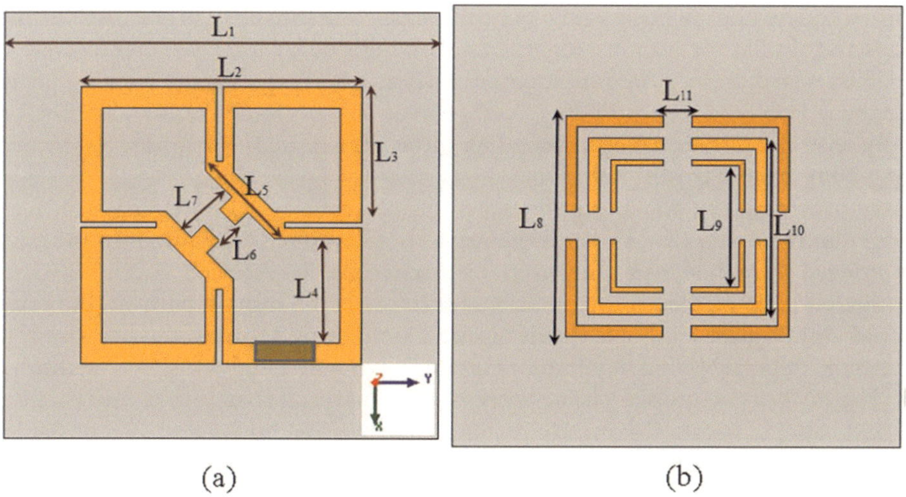

Figure 13.
Geometry of the antenna: (a) top view, and (b) MMT load at the bottom layer [41].

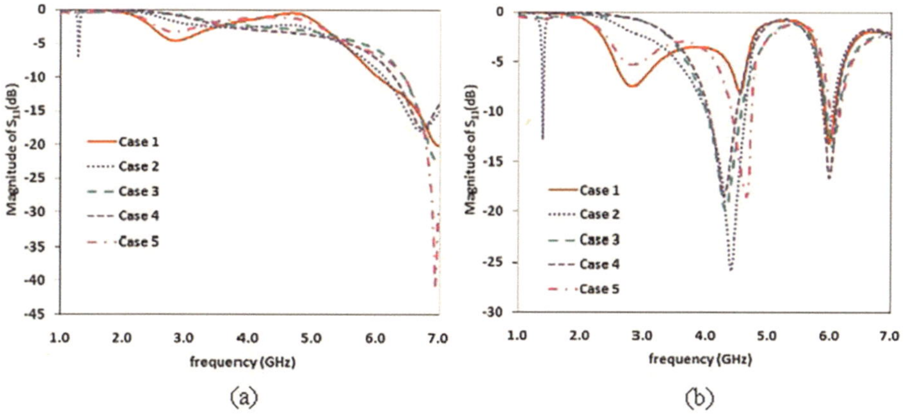

Figure 14.
Reflection coefficient (S_{11}): (a) without metamaterial, and (b) with metamaterial [41].

without metamaterial is presented in **Figure 14**. The S_{11} shows that the proposed antenna operates at 7 GHz (**Figure 14**). But when a metamaterial layer of multi parallel rings are added, the resonance frequency is shifted from 7 GHz to 4 GHz and the size of the antenna is considerably reduced.

3.1.4 Metamaterials for the reduction of specific absorption rate

Wireless body area networks (WBAN) are widely employed in many different applications, including mobile communication, military applications, medical diagnosis, and rescue services. In these kinds of applications, the antennas must be operated in close contact to the human body. The backward radiation is absorbed into human body tissues as a result of its near vicinity to the human body and may seriously harm these tissues. The absorption of energy by human body tissue is defined by the specific absorption rate (SAR) [42]. The SAR is defined as the rate of electromagnetic energy absorbed by the human body tissue per unit mass. The safe limit of the SAR for mobile phones and similar electronic gadgets has been defined by international standards. The US standard defines the safe limit for SAR as 1.6 W/kg average over 1 g of tissue, while the EU defined it as 2.0 W/kg average over 10 g of tissue [27, 43]. The SAR must comply with the safe limits established by international bodies in order to protect the human body from harmful radiation.

Several techniques, including the use of reflectors [44], RF shielding with ferrite and conductive materials [45], and the use of highly directional antennas [46], have been utilized to reduce SAR. Recently, metamaterials like AMC, SRR, and EBG are investigated to block electromagnetic waves towards the human body and to reduce SAR [47–49]. **Figure 15** shows the design and simulation setup of the proposed antennas with polarization dependent metamaterial and a high effective medium ratio [50]. The SAR of the mobile phone without metamaterial structure is calculated and analyzed using L, S, and C-band frequencies, considering both 1 and 10 g of tissues. In this case, the metamaterial structure acts as a shielding material and protects the human head from harmful radiation. It is evident that the unique metamaterial structure lowers the SAR value emitted from a mobile phone by 98%. Likewise, a metamaterial based fabric antenna at ISM band is shown in **Figure 16**, where the metamaterial structure behaves as EBG/AMC [51]. The proposed antenna is analyzed for SAR with and without metamaterial structures.

The SAR analysis of the traditional and the antenna integrated with metamaterial at the ISM band considering the Federal Communications Commission (FCC) standard of 1 g of human body tissue is presented in **Figure 17**. It is observed that the traditional antenna, when placed in the vicinity of human body, gives a SAR of 7.78 W/kg, exceeding the safe limit. However, the antenna integrated with metamaterial when placed in close contact with the human body gives an SAR value of 0.028 W/kg, which is within the safer limit of the FCC standard of 1.6 W/kg averaged

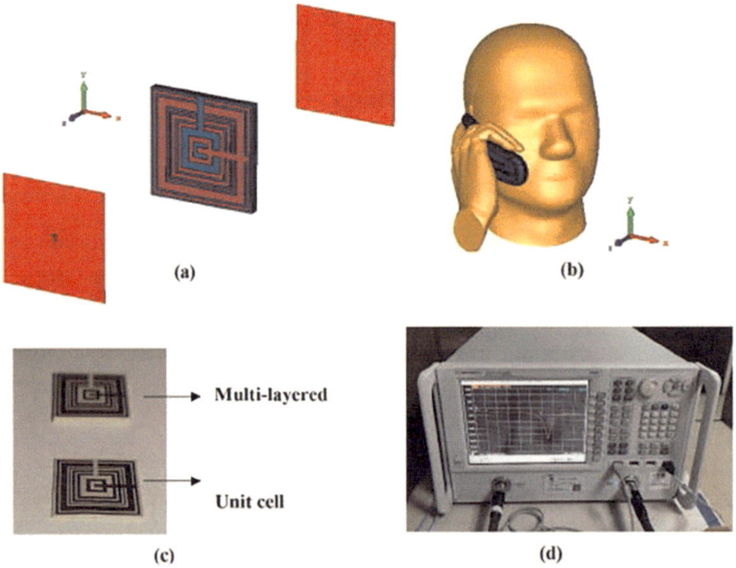

Figure 15.
Layout and simulation setup: (a) metamaterial structure, (b) Voxel model (human head and hand), (c) Fabricated metamaterial structure, (d) Agilent N5227A VNA [50].

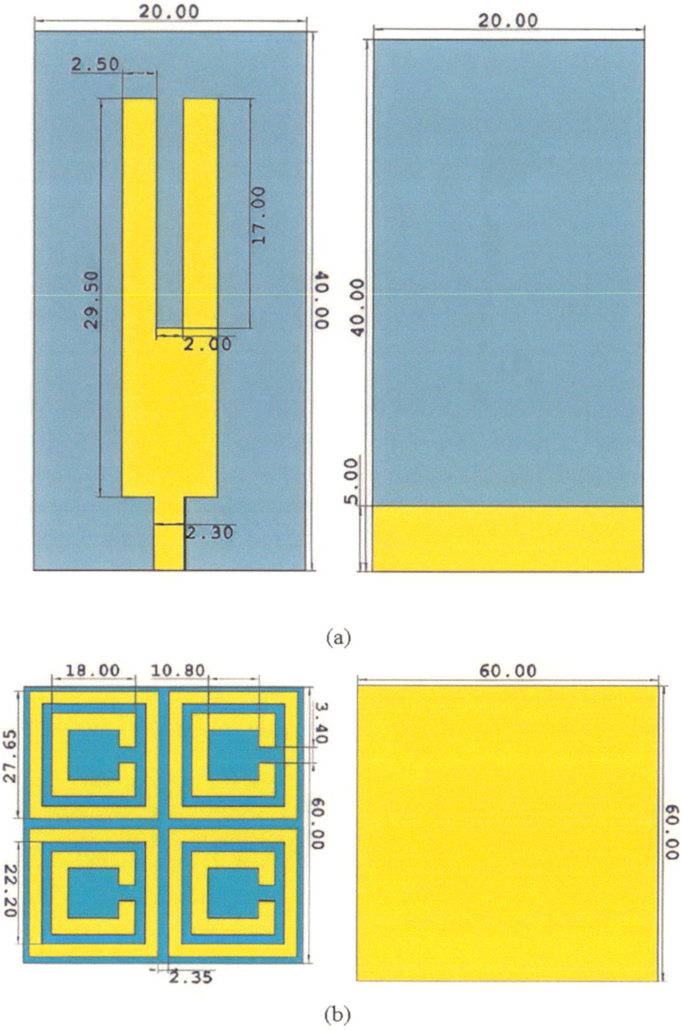

Figure 16.
Design layout for: (a) U-shaped fabric antenna, and (b) Metamaterial on fabric material [51].

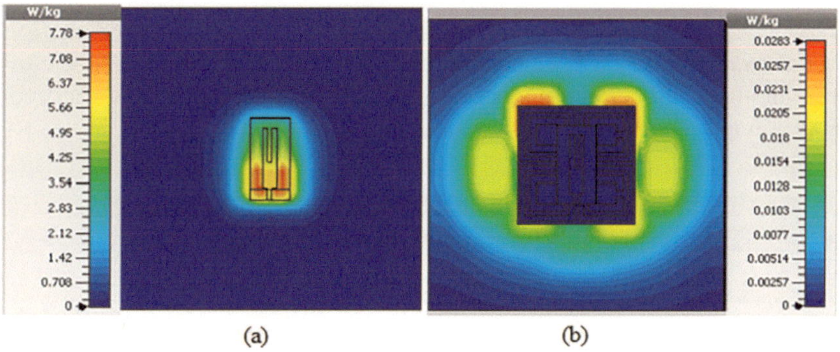

Figure 17.
Simulated SAR of the proposed U-shaped antenna (a) without metamaterial, and (b) with metamaterial [51].

over 1 g of tissue. Hence, the metamaterial integrated antenna play a vital role to reduce the SAR significantly compared to its conventional counterpart.

4. Applications of metasurfaces

4.1 Metasurfaces for polarization conversion

Polarization is the characteristic of a wave's oscillation, i.e., having a particular direction relative to the wave's propagation. The polarization of an EM wave is defined as the direction of the electric field oscillation in a plane transverse to the propagation. The process of altering the polarization state of a wave after reflecting, transmitting from a medium is known as polarization conversion. Wave plates, also known as retarders, modify the polarization state by retarding (or delaying) one component of polarization with respect to its orthogonal component. Wave plates alter the polarization state of EM waves. The most common types of wave plates include the half-wave plates and quarter-wave plates. The former changes the path of linearly polarized EM waves, while the latter transforms linearly polarized EM waves to circular polarized one, and vice versa. Every polarization state can be decomposed into orthogonal components, and the phase difference between them must be regulated in order to convert one polarization to another. Metasurface structures with asymmetric resonators having low co-polarization reflection and high cross-polarization reflection are usually used for polarization conversion. In the following subsections, the mathematical background and designs examples of the polarization converting metasurfaces will be discussed.

4.1.1 Polarization converting metasurface

Polarization converting metasurface (PCMS) is a device that transforms the polarization state of an incident EM wave with no significant loss. When an incident EM wave is transformed from y-polarized to x-polarized wave upon reflection from the metasurface, the corresponding conversion is called cross polarization conversion. Generally, the reflection coefficient is defined as a ratio of reflected field E^{ref} and incident field E^{inc}, (i.e., $r = E^{\text{ref}}/E^{\text{inc}}$). The electric field is a plane wave propagating along the z-direction and is given as:

$$E = \hat{u}^{\text{pol}} E_o e^{-j(kz-\omega t)} \tag{5}$$

where $\hat{u}^{\text{pol}}$, E_o, k, and ω, represents, respectively, the polarization direction, amplitude of electric field, wave number, and the angular frequency. The correlation among the incident and reflected electric fields of LP incidence wave oriented along $(x\text{–}y)$-direction is given by the Jones matrix [52]:

$$\begin{bmatrix} E_x^{\text{ref}} \\ E_y^{\text{ref}} \end{bmatrix} = \begin{bmatrix} R_{xx} & R_{xy} \\ R_{yx} & R_{yy} \end{bmatrix} \begin{bmatrix} E_x^{\text{inc}} \\ E_y^{\text{inc}} \end{bmatrix} \tag{6}$$

where R_{xx} and R_{yy} represents co-polarized reflection and are defined as $R_{xx} = E_x^{\text{ref}}/E_x^{\text{inc}}$, $R_{yy} = E_y^{\text{ref}}/E_y^{\text{inc}}$, similarly R_{yx} and R_{xy} represents cross-polarized reflection and are defined as $R_{yx} = E_y^{\text{ref}}/E_x^{\text{inc}}$, $R_{xy} = E_x^{\text{ref}}/E_y^{\text{inc}}$. E_x^{inc}, E_x^{ref} is the incident and reflected x-polarized electric field and vice versa, respectively.

For efficient cross polarization transformation, the level of the cross-polarized reflection coefficients should be larger than −3 dB, i.e., $|R_{yx}|,|R_{xy}| >$ -3 dB while the level of co-polarized reflection coefficient should be smaller than −10 dB, i.e., $|R_{xx}|$, $|R_{yy}| < -10$ dB [52]. To illustrate how polarization transformation is achieved, a metasurface composed of an arrow shaped resonator is designed as shown in **Figure 18(a)**. It is seen from the results shown in **Figure 19(a)** that the proposed PCMS fulfilled the aforementioned criteria and performs cross polarization conversion (CPC) in an ultra-wideband (10.1–26.0 GHz). Polarization conversion ratio (PCR) is usually expressed as the ratio of the power reflected in the cross-polarized component to the total reflected power, i.e., [53]:

$$\text{PCR} = \frac{|R_{yx}^2|}{|R_{yx}^2| + |R_{xx}^2|}. \tag{7}$$

The PCR results presented in **Figure 19(b)** show that the PCR value is greater than 90% from 10.1 to 26 GHz, indicating that most of the energy is converted to its orthogonal counter part in this band. The R_{xx} attains three minimum values at: −49.6, −54.5, −45.3 dB, and the corresponding frequencies are 11.1, 17.5, and 25.1 GHz, respectively with a PCR of 100% at 11.1, 17.5, and 25.1 GHz frequencies.

The incident wave with polarization direction along the y-axis is decomposed into u- and v-components to explain the cross polarization conversion mechanism, as illustrated in **Figure 18(b)**. The u- and v components are mutually orthogonal and

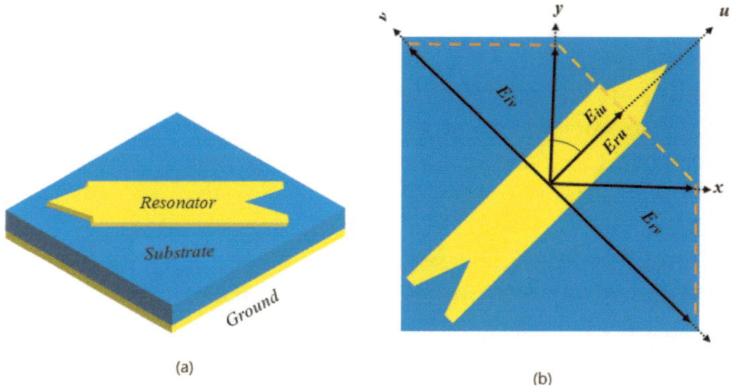

Figure 18.
The designed PCMS: (a) unit cell, and (b) u − v decomposition.

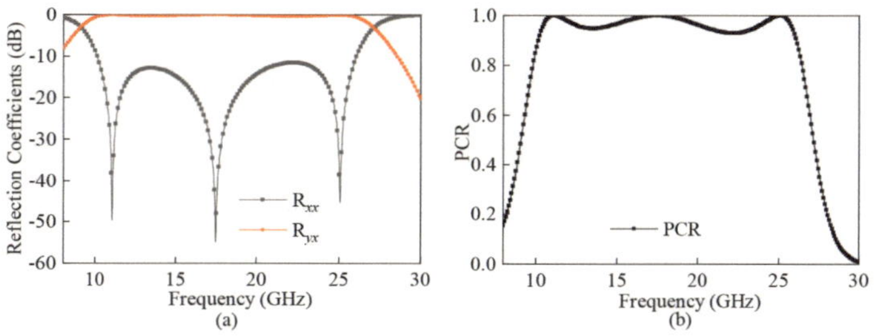

Figure 19.
Simulated reflection coefficients of LP waves: (a) x, and (b) PCR of the proposed metasurface.

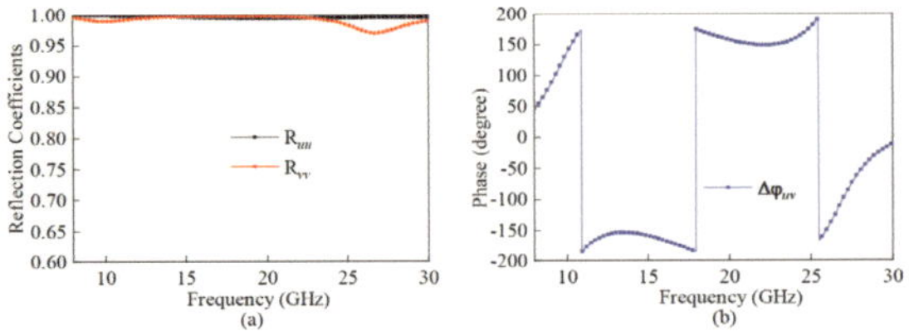

Figure 20.
Simulated reflection coefficients: (a) u and v, and (b) Phase difference.

form a ±45° angle with the y-axis. The y-polarized incident electric field $\overrightarrow{E_{\mathrm{i}}}$, along the u- and v-axis is expressed as:

$$\overrightarrow{E_{\mathrm{i}}} = \hat{u}E_{\mathrm{i}u}e^{j\phi} + \hat{v}E_{\mathrm{i}v}e^{j\phi} \tag{8}$$

where $\hat{u}$ and $\hat{v}$ are the unit vectors, and φ is the phase of the components. The x-polarized reflected electric field, i.e., $\overleftarrow{E_{\mathrm{r}}}$, can be expressed as:

$$\overrightarrow{E_{\mathrm{r}}} = \hat{u}E_{\mathrm{r}u}e^{j\phi} + \hat{v}E_{\mathrm{r}v}e^{j(\phi \pm 180)}, \tag{9}$$

where $\varphi = \varphi_{uu}$ and $\varphi \pm 180° = \varphi_{vv}$. The co-polarized reflection coefficients along the u- and v-axes are $r_{uu} = E_u^{\mathrm{ref}}/E_u^{\mathrm{inc}}$ and $r_{vv} = E_v^{\mathrm{ref}}/E_v^{\mathrm{inc}}$. The reflection amplitudes are ($r_{uu} = r_{vv} = 1$) and the phase difference is ($\Delta\varphi = \varphi_{uu} - \varphi_{vv} = \pm 180°$) for CPC through reflected wave in the u- and v-direction [54].

The unit cell was simulated to polarized incidences in the u- and v directions to test the CPC. The results are shown in **Figure 20**. **Figure 20(a)** shows that the amplitudes of the co-polarized coefficients r_{uu} and r_{vv} are close to unity. **Figure 20(b)** shows that in the frequency range (10.1–26 GHz), the phase difference between the u- and v-axis components does not vary much over 180°, validating the effectiveness of cross polarization conversion.

For reflection type of PCMS, the surface currents on both top and bottom surfaces are found to be diagonal. As a result, the field components along the original incident direction are canceled whereas the other one is enhanced in the mutual orthogonal direction. Because the induced field is oriented in an orthogonal direction, it can produce a perpendicularly oriented electric field to the incident one, resulting in polarization conversion of EM waves [55].

The principle of CPC is further explained by the surface current distributions shown in **Figure 21**. At resonance frequecies of 11.1 and 17.6 GHz, the surface currents on the patch and ground are anti parallel resulting in magnetic resonance. At resonance frequency if 25.1 GHz, the surface currents on the patch and ground are parallel resulting in electric resonance. The occurrence of electric and magnetic resonances results in wideband polarization conversion with a high PCR.

4.2 Metasurfaces for radar cross section reduction

Radar cross section (RCS), an essential property of a radar, is a measure of how detectable an object is by radar. The larger the RCS, the easier it is for the object to be detected. A unit cell and its mirror are required for the RCS reduction using polarization converting technique. The unit cell and its mirror unit cell has a cross polarization

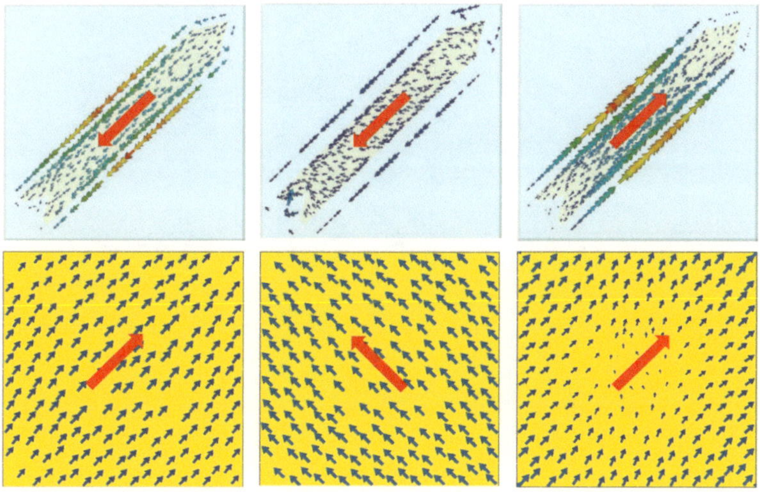

Figure 21.
Surface current distribution of the top metal and ground metal at resonant eigen-modes: (a) 11.1 GHz, (b) 17.6 GHz, and (c) 25.1 GHz.

phase difference of 180°, which fully satisfies the requirement for RCS reduction. For RCS reduction, chessboard structures are utilized that usually use single PCMS unit and its mirror unit (rotating the same structure by 90°). As per the polarization conversion concept, the phase difference between the PCMS and its mirror should always be 180°. As a result, the PCR is the only component that influences RCS reduction performance. So, when an x-polarized incident waves is reflected, it totally changes to an orthogonal counterpart. The PCMS units reflections, as well as their mirror reflections, are canceled out. As a result, by arranging PCMS units and their mirrors in chessboard structures, the RCS can be minimized.

Figure 22(a) shows the reflection coefficients of the mirror PCMS, which are the same as the reflection coefficients of PCMS. **Figure 22(b)** plots the phase difference between the reflection phases of PCMS and its mirror PCMS unit cells. In the whole PCR bandwidth, there is a 180 reflection phase difference between PCMS and its mirrored PCMS unit cell, indicating that the proposed PCMS satisfies the reflection

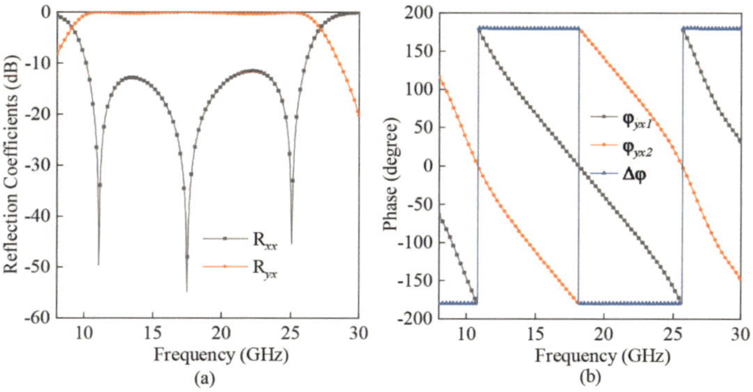

Figure 22.
Simulated results of: (a) reflection coefficients of mirror PCM, and (b) reflection phase of PCM and its mirror.

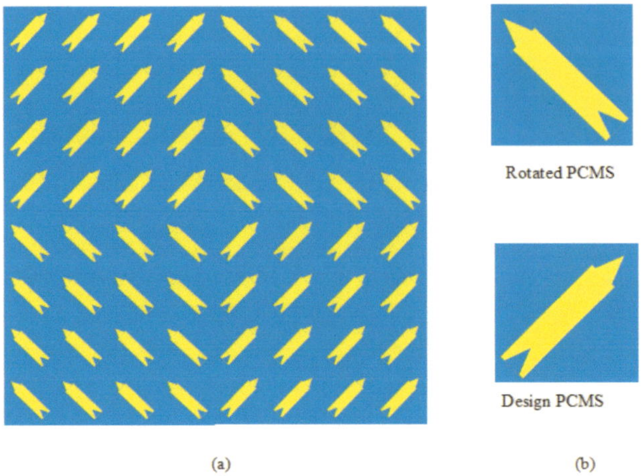

Figure 23.
Top view of the designed square chessboard array with PCMS and its mirror.

phase cancelation requirements for RCS reduction. The RCS reduction from the surface compared to the PEC can be expressed as [56]

$$\text{RCS reduction (dB)} = 10\log_{10}\left[\frac{\lim_{r\to\infty}4\pi r^2\frac{|\vec{E}_{rx}|^2}{|\vec{E}_{ix}|^2}}{\lim_{r\to\infty}4\pi r^2|1|^2}\right] \\ = R_{xx}\,(\text{dB}) = 10\log_{10}(1-\text{PCR}), \tag{10}$$

where $|E_{rx}|$ and $|E_{ix}|$ are the reflected and incident fields in the far zone of $r \to \infty$, respectively, r is the detecting distance, and R_{xx} represents the co-polarized reflection coefficients [56]. According to the above equation, RCS reduction is related to PCR. The bandwidth of RCS reduction is predicted from bandwidth of PCR. If PCR is greater than 90%, a 10-dB RCS reduction is achieved.

Based on the designed PCMS unit, an array with 8 × 8 units and a total dimension of 56 × 56 was designed in square chessboard configuration as shown in **Figure 23**. The designed PCMS is rotated at an angle of 90, 180, and 270° to make a square chessboard configuration, which reflects the wave in-phase and out-phase to create destructive interference, resulting in RCS reduction. The comparison of the proposed square chessboard array monostatic RCS performance with a PEC of the same dimension is shown in **Figure 24**. The designed square chessboard array achieved ultra-wideband RCS reduction in a frequency band from 9.6 to 27.5 GHz. The RCS reduction is maximum at the resonance frequencies of 10.3, 11.8, 19.8, and 20.3 GHz, with the maximum RCS reduction of 27.8 dB observed at 10.3 GHz.

4.3 Metasurfaces for absorption

Metasurfaces can also be utilized for designing perfect absorbers, which absorb most energy of the incident EM wave within the desired frequency band. Metasurface structures with symmetric resonators have been used as effective absorbers with low

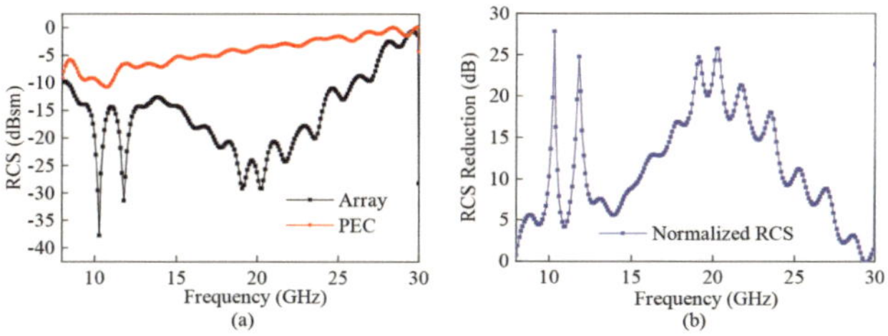

Figure 24.
Mono-static RCS reduction: (a) PCMS and PEC, and (b) normalized RCS.

co- and cross polarization reflection [57]. Metasurface absorbers (MSAs) usually consists of subwavelength resonant metal insulator metal (MIM) structures. Mathematically, the absorption can be expressed by the following equation

$$A(\omega) = 1 - R(\omega) - T(\omega), \tag{11}$$

where

$$R(\omega) = \frac{Z_\omega - Z_o}{Z_\omega - Z_o}, \tag{12}$$

$$Z(\omega) = \sqrt{\frac{\mu_o \mu_r(\omega)}{\varepsilon_o \varepsilon_r(\omega)}}, \tag{13}$$

$$Z_0 = \sqrt{\frac{\mu(\omega)}{\varepsilon(\omega)}}, \tag{14}$$

with Z_o being the characteristic impedance of the free space, which is equal to 377 Ω. The free space permeability and permittivity are denoted by μ_o, and ε_o, and μ_r, ε_r represents the relative permeability and permittivity in the material. To design an MSA, the unit cell's parameters should be precisely adjusted such that the absorbers impedance matches the free space impedance so the reflection is minimized, resulting in perfect absorption of the incident waves. The effective impedance of the MSA matches the free space impedance by altering the electric and magnetic resonance simultaneously. Metasurface absorbers are characterized by unit cells having single and multi resonant designs, transforming into narrowband and broadband absorption, respectively [58]. In general, the resonance frequency of the MSA is determined by the geometric configuration and geometric characteristics of a unit cell. Metasurface based absorbers are low profile, light weight, with simple metallic structures. They have limitations in terms of absorption bandwidth, absorption rate and air breakdown at high power [59]. The bottom side of the metasurface absorber is a metallic sheet. It acts as a ground, which causes no transmission and consequently $T(\omega)$ is equal to zero. Absorption can also be defined as:

$$A(\omega) = 1 - |R_{xx}|^2 - |R_{yx}|^2 - T, \tag{15}$$

where $|R_{xx}|^2$ and $|R_{yx}|^2$ represents the co- and cross-polarized reflected powers, the transmitted power, i.e., T, is zero because of the metallic ground plane. For absorption both components of the reflected wave, i.e., co and cross should be zero. A metasurface absorber is designed using square ring resonator made of copper placed at the top of dielectric substrate FR-4 having metallic ground plane on the back side as shown in **Figure 25**. The plots of reflection coefficients and absorption are shown in **Figure 26**. For perfect absorption, the levels of both the co- and cross components of the reflection coefficients should be smaller than −10 dB, i.e., $|R_{xx}|, |R_{yx}| < -10$ dB. The results shown in **Figure 26(a)** indicates that both the co- and cross components of the reflection coefficients are below −10 dB, which indicates that all of the incident wave is absorbed. The results in **Figure 26(b)** show that absorption remains above 90%, which is requirement for perfect absorption.

The intrinsic parameters of the metasurface, i.e., effective permittivity (ε_{eff}), effective permeability (μ_{eff}), and effective impedance (z), are obtained by S-parameter-retrieval method [60, 61]. These parameters are extracted as:

$$z = \pm\sqrt{\frac{(1+S_{11})^2 - S_{21}^2}{(1+S_{11})^2 - S_{21}^2}}, \tag{16}$$

$$e^{jnK_oH} = \frac{S_{21}}{1 - S_{11}\left(\frac{z-1}{z+1}\right)}, \tag{17}$$

$$\varepsilon_{eff} = \frac{n}{z}, \tag{18}$$

$$\mu_{eff} = n \times z, \tag{19}$$

Figure 25.
Proposed metasurface absorber.

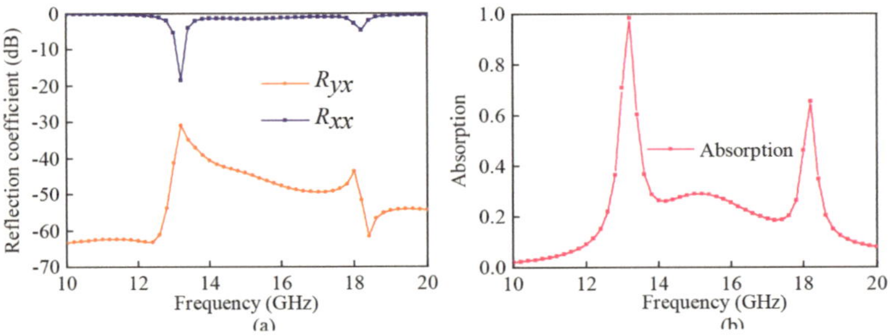

Figure 26.
(a) Reflection coefficients, and (b) absorption.

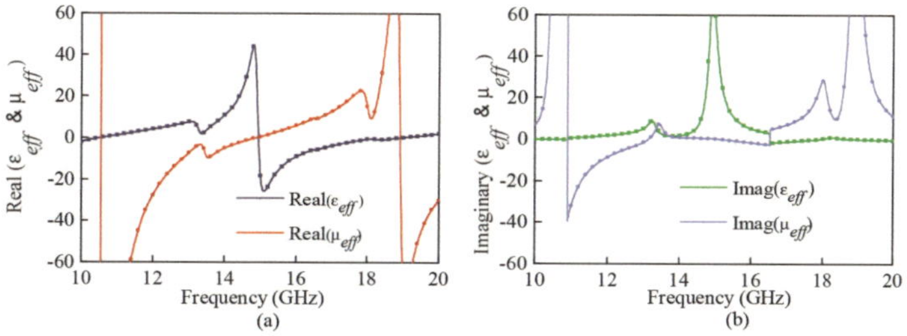

Figure 27.
Extracted effective parameters: (a) real, and (b) imaginary parts.

where z stands for impedance, S_{11} and S_{21} for reflection and transmission amplitudes, n for refractive index, K_o for wave number, and H for substrate thickness [62]. The values of the extracted parameters are shown in **Figure 27**. At peak frequencies of 7.8 and 16.7 GHz, the real and imaginary components of the extracted permittivity, and permeability obtained are almost equal, resulting in impedance matching of the MS with free space, which minimizes the co-polarization reflection and maximizes the absorption [63].

5. Conclusions

This chapter described the applications of metamaterials and metasurfaces. The underlying principles and design considerations of metamterials in antenna design were explained in detail. The state-of-the-art works related to the application of metamaterials in antenna designs were reviewed. Also it is discussed that how metamaterials are used to improve and enhance the antenna parameters such as gain, directivity, bandwidth, size, and SAR. Moreover, the theoretical concepts and mathematical analysis of metasurfaces for polarization conversion, RCS reduction, and absorption were also discussed. Some metasurfaces were designed to show how to achieve different desired functionalities.

Author details

Babar Kamal[1*], Usman Ali[2], Jingdong Chen[1] and Sadiq Ullah[2]

1 Center of Intilligent Acoustics and Immersive Communication, School of Marine Science and Technology, Northwestern Polytechnical University, Xian, China

2 University of Engineering and Technology, Mardan, Pakistan

*Address all correspondence to: babarkamal.55@mail.nwpu.edu.cn

References

[1] Shivola A. Metamaterials in electromagnetics. Metamaterials. 2007; **1**(1):2-11. DOI: 10.1016/j.metamat.2007.02.003

[2] Smith DR, Padilla WJ, Vier DC, Nemat-Nasser SC, Schultz S. Composite medium with simultaneously negative permeability and permittivity. Physical Review Letters. 2000;**84**(18):4184. DOI: 10.1103/PhysRevLett.84.4184

[3] Walia S, Shah CM, Gutruf P, et al. Flexible metasurfaces and metamaterials: A review of materials and fabrication processes at micro-and nano-scales. Applied Physics Review. 2015;**2**(1):011303. DOI: 10.1063/1.4913751

[4] Freire MJ, Marques R, Jelinek L. Experimental demonstration of a Âμ = −1 metamaterial lens for magnetic resonance imaging. Applied Physics Letters. 2008;**93**(23):1-4. DOI: 10.1063/1.3043725

[5] Ozbey B, Demir HV, Kurc O, Erturk VB, Altintas A. Wireless measurement of elastic and plastic deformation by a metamaterial-based sensor. Sensors. 2014;**14**(10):19609-19621. DOI: 10.3390/s141019609

[6] Liu W, Chen ZN, Qing X. Metamaterial-based low-profile broad-band mushroom antenna. IEEE Transactions on Antennas and Propagation. 2013;**62**(3):1165-1172. DOI: 10.1109/TAP.2013.2293788

[7] Glybovski SB, Tretyakov SA, Belov PA, Kivshar YS, Simovski CR. Metasurfaces: From microwaves to visible. Physics Reports. 2016;**634**:1-72. DOI: 10.1016/j.physrep.2016.04.004

[8] Holloway CL, Kuester EF, Gordon JA, OHara J, Booth J, Smith DR. An overview of the theory and applications of metasurfaces: The two-dimensional equivalents of metamaterials. IEEE Antennas Propagation Magzines. 2012; **54**:10-35. DOI: 10.1109/MAP.2012.6230714

[9] Chen HT, Taylor AJ, Yu N. A review of metasurfaces: Physics and applications. Reports on Progress in Physics. 2016;**79**(7):076401. DOI: 10.1088/0034-4885/79/7/076401

[10] Li F, Chen H, He Q, Zhou Y, Zhang L, Weng X, et al. Design and implementation of metamaterial polarization converter with the reflection and transmission polarization conversion simultaneously. Journal of Optics. 2020;**21**(4):04510. DOI: 10.1088/2040-8986/ab0878

[11] Roy T, Rogers ET, Zheludev NI. Sub-wavelength focusing meta-lens. Optics Express. 2013;**21**(6):75777582

[12] Azad AK et al. Metasurface broad-band solar absorber. Scientific Reports. 2016;**6**(1):16. DOI: 10.1038/srep20347

[13] Ni X et al. An ultrathin invisibility skin cloak for visible light. Science. 2015; **349**(6254):13101314. DOI: 10.1126/science.aac9411

[14] Zhang XG et al. A metasurface-based light-to-microwave transmitter for hybrid wireless communication. Light: Science & Applications. 2022;**11**(126):1-10. DOI: 10.1038/s41377-022-00817-5

[15] Khan B, Kamal B, Ullah S, Abdullah AH, Ullah R. Asymmetric polarization converting metasurface fo micowave application. Optical Material Express. 2022;**12**(9):3403-3415. DOI: 10.1364/OME.455472

[16] Amiri M, Tofigh F, Shariati N, Lipman J, Abolhasan M. Review on metamaterial perfect absorbers and their applications to IoT. IEEE Internet of Things Journal. 2020;**8**(6):4105-4131. DOI: 10.1109/JIOT.2020.3025585

[17] Milias C, Andersen RB, Lazaridis PI, Zaharis ZD, Muhammad B, Kristensen JT, et al. Metamaterial-inspired antennas: A review of the state of the art and future design challenges. IEEE Access. 2021;**9**:89846-89865. DOI: 10.1109/ACCESS.2021.3091479

[18] Ali Esmail B, Majid HA, Zainal Abidin Z, Haimi Dahlan S, Himdi M, Dewan R, et al. Reconfigurable radiation pattern of planar antenna using metamaterial for 5G applications. Materials. 2020;**13**(3):582. DOI: 10.3390/ma13030582

[19] Agarwal S, Prajapati YK. Multifunctional metamaterial surface for absorbing and sensing applications. Optics Communications. 2019;**439**:304-307. DOI: 10.1016/j.optcom.2019.01.020

[20] Tadesse AD, Acharya OP, Sahu S. Application of metamaterials for performance enhancement of planar antennas: A review. International Journal of RF and Microwave Computer & Aided Engineering. 2020;**30**(5):e22154. DOI: 10.1002/mmce.22154

[21] Asif M, Sehrai DA, Kiani SH, Khan J, Abdullah M, Ibrar M, et al. Design of a dual band SNG metamaterial based antenna for LTE 46/WLAN and Ka-band applications. IEEE Access. 2021;**9**: 71553-71562. DOI: 10.1109/ACCESS.2021.3077844

[22] Rosaline SI. A triple-band antenna with a metamaterial slab for gain enhancement and specific absorption rate (SAR) reduction. Progress in Electromagnetic Research C. 2021;**109**: 275-287. DOI: 10.2528/PIERC20122202

[23] Ali U, Ullah S, Shafi M, Shah SA, Shah IA, Flint JA. Design and comparative analysis of conventional and metamaterial?based textile antennas for wearable applications. International Journal of Numerical Modelling: Electronic Networks, Devices and Fields. 2019;**32**(6):e2567. DOI: 10.1002/jnm.2567

[24] Yayun D, Wenwen L, Xijun Y, Chen Y, Houjun T. Design of unit cell for metamaterials applied in a wireless power transfer system. In: IEEE PELS Workshop on Emerging Technologies: Wireless Power Transfer. 2017. pp. 143–147. DOI: 10.1109/WoW.2017.7959382

[25] Phan DT, Phan HL, Nguyen TQH. A miniaturization of microstrip antenna using negative permittivity metamaterial based on CSRR loaded ground for WLAN applications. Journal of Science and Technology. 2016;**54**(6):689-697. DOI: 10.15625/0866-708X/54/6/8375

[26] Krzysztofik WJ, Cao TN. Metamaterials in application to improve antenna parameters. Metamaterials and Metasurfaces. 2018;**12**(2):63-85

[27] Ali U, Ullah S, Khan J, Shafi M, Kamal B, Basir A, et al. Design and SAR analysis of wearable antenna on various parts of human body, using conventional and artificial ground planes. Journal of Electrical Engineering and Technology. 2017;**12**(1):317-328. DOI: 10.5370/JEET.2017.12.1.317

[28] Lu G, Yan F, Zhang K, Zhao Y, Zhang L, Shang Z, et al. A dual-band high-gain subwavelength cavity antenna with artificial magnetic conductor metamaterial microstructures. Micromachines. 2021;**13**(1):58. DOI: 10.3390/mi13010058

[29] Gong Y, Yang S, Li B, Chen Y, Tong F, Yu C. Multi-band and high gain antenna using AMC ground characterized with four zero-phases of reflection coefficient. IEEE Access. 2020; **8**:171457-171468. DOI: 10.1109/ACCESS.2020.3024982

[30] Sievenpiper D, Zhang L, Broas RF, Alexopolous NG, Yablonovitch E. High-impedance electromagnetic surfaces with a forbidden frequency band. IEEE Transactions on Microwave Theory and Techniques. 1999;**47**(11):2059-2074. DOI: 10.1109/22.798001

[31] Yanagi T, Oshima T, Oh-hashi H, Konishi Y, Murakami S, Itoh K, et al. Lumped-element loaded EBG structure with an enhanced bandgap and homogeneity. In: 2008 International Workshop on Antenna Technology: Small Antennas and Novel Metamaterials. 2008. pp. 458-461. DOI: 10.1109/IWAT.2008.4511377

[32] Zheng YJ, Gao J, Zhou YL, Cao XY, Xu LM, Li SJ, et al. Metamaterial based patch antenna with wideband RCS reduction and gain enhancement using improved loading method. IET Microwaves, Antennas & Propagation. 2017;**11**(9):1183-1189. DOI: 10.1049/iet-map.2016.0746

[33] Roy S, Chakraborty U. Metamaterial-embedded dual wideband microstrip antenna for 2.4 GHz WLAN and 8.2 GHz ITU band applications. Waves in Random and Complex Media. 2020;**30**(2):193-207. DOI: 10.1080/17455030.2018.1494396

[34] Sumathi K, Lavadiya S, Yin P, Parmar J, Patel SK. High gain multiband and frequency reconfigurable metamaterial superstrate microstrip patch antenna for C/X/Ku-band wireless network applications. Wireless Networks. 2021;**27**(3):2131-2146. DOI: 10.1007

[35] Jeong H, Kim Y, Tentzeris MM, Lim S. Gain-enhanced metamaterial absorber-loaded monopole antenna for reduced radar cross-section and back radiation. Materials. 2020;**13**(5):1247. DOI: 10.3390/ma13051247

[36] Bakhtiari A, Sadeghzadeh RA, Moghadasi MN. Gain enhanced miniaturized microstrip patch antenna using metamaterial superstrates. IETE Journal of Research;**65**(5):635-640. DOI: 10.1080/03772063.2018.1447406

[37] Gao XJ, Cai T, Zhu L. Enhancement of gain and directivity for microstrip antenna using negative permeability metamaterial. AEU-International Journal of Electronics and Communications. 2016;**70**(7):880-885. DOI: 10.1016/j.aeue.2016.03.019

[38] Khoutar FZ, Aznabet M, Mrabet OEL, Nayat-Ali O. Double layer metamaterial superstrate for performance enhancement of a microstrip patch antenna. In: ICCWCS 2019: Third International Conference on Computing and Wireless Communication Systems, ICCWCS 2019, April 24-25, Faculty of Sciences, Ibn Tofal University-Knitra-Morocco, 82. European Alliance for Innovation DOI: 10.4108/eai.24-4-2019.2284232

[39] Al-Bawri SS, Islam MS, Wong HY, Jamlos MF, Narbudowicz A, Jusoh M, et al. Metamaterial cell-based superstrate towards bandwidth and gain enhancement of quad-band CPW-fed antenna for wireless applications. Sensors. 2020;**20**(2):457. DOI: 10.3390/s20020457

[40] Raval F, Kosta YP, Joshi H. Reduced size patch antenna using complementary split ring resonator as defected ground

plane. AEU-International Journal of Electronics and Communications. 2015; **69**(8):1126-1133. DOI: 10.1016/j. aeue.2015.04.013

[41] Varamini G, Keshtkar A, Naser-Moghadasi M. Miniaturization of microstrip loop antenna for wireless applications based on metamaterial metasurface. AEU-International Journal of Electronics and Communications. 2018;**83**:32-39. DOI: 10.1016/j. aeue.2017.08.024

[42] Hediya AM. Reduction of specific absorption rate: A review article. The Egyptian International Journal of Engineering Sciences and Technology. 2022. DOI: 10.21608/eijest.2022. 108455.1117

[43] Karimian R, Ardakani MD, Ahmadi S, Zaghloul M. Human body specific absorption rate reduction employing a compact magneto-dielectric AMC structure for 5G massive-MIMO applications. Engineering. 2021;**2**(4): 501-511. DOI: 10.3390/eng2040032

[44] Haridim M. Use of rod reflectors for SAR reduction in human head. IEEE Transactions on Electromagnetic Compatibility. 2015;**58**(1):40-46. DOI: 10.1109/TEMC.2015.2500818

[45] Stephen JP, Hemanth DJ. An investigation on specific absorption rate reduction materials with human tissue cube for biomedical applications. International Journal of RF and Microwave Computer Aided Engineering. 2019;**29**(12):e21960. DOI: 10.1002/mmce.21960

[46] Pikale R, Sangani D, Chaturvedi P, Soni A, Munde M. A review: Methods to lower specific absorption rate for mobile phones. In: 2018 International Conference On Advances in Communication and Computing Technology (ICACCT). IEEE. 2018. pp. 340-343. DOI: 10.1109/ICACCT.2018. 8529654

[47] Sugumaran B, Balasubramanian R, Palaniswamy SK. Reduced specific absorption rate compact flexible monopole antenna system for smart wearable wireless communications. Engineering Science and Technology, an International Journal. 2021;**24**(3):682-693. DOI: 10.1016/j.jestch.2020.12.012

[48] Imaculate Rosaline S, Raghavan S. Design and analysis of a SRR superstrate for SAR reduction. Journal of Electromagnetic Waves and Applications. 2015;**29**(17):2330-2338. DOI: 10.1080/09205071.2015.1091384

[49] Tamim AM, Faruque MRI, Khandaker MU, Islam MT, Bradley DA. Electromagnetic radiation reduction using novel metamaterial for cellular applications. Radiation Physics and Chemistry. 2021;**178**:108976. DOI: 10.1016/j.radphyschem.2020.108976

[50] Tayaalen R, Mohammad RIF, Mohammad TI. Specific absorption rate reduction for sub-6 frequency range using polarization dependent metamaterial with high effective medium ratio. Scientifc Reports. 2022;**12** (1):1-18. DOI: DOI; 10.1007/s11664-019-07156-z

[51] Ashyap AY, Zainal Abidin Z, Dahlan SH, Majid HA, Saleh G. Metamaterial inspired fabric antenna for wearable applications. International Journal of RF and Microwave Computer Aided Engineering. 2019;**29**(3):e21640. DOI: 10.1002/mmce.21640

[52] Kamal B, Chen J, Yin Y, Ren J, Ullah S, Khan B. Broad band and broad angle linear ans circular polarization converting metasurfce. Journal of Electromagnetic Waves and

Applications. 2022;**36**(8):1102-1112. DOI: 10.1080/09205071.2021.2005696

[53] Kamal B, Chen J, Yin Y, Ren J, Ullah S, Ali U. Design and experimental analysis of dual-band polariza- tion converting metasurface. IEEE Antennas Wireless Propagation Letter. 2021;**20**(8): 1409-1413. DOI: 10.1109/LAWP.2021.3083334

[54] Abdullah KB, Ullah S, Khan B, Ali H, Rahim T. Cross polarization conversion metasurface for fixed wireless and ku-band applications. International Journal of Communication Systems. 2022;**35**(3): e5033. DOI: 10.1002/dac.5033

[55] Bhattacharya S, Saha C, Siddiquiand JY. High frequency applications of metamaterials and metasurfaces. In: Proceedings of the IEEE Recent Advances in Geoscience and Remote Sensing: Technologies, Standards and Applications (TENGARSS'19); 17-20 October 2019: Kochi. India IEEE. 2019. pp. 96–99

[56] Fu C, Han L, Liu C, Sun Z, Lu X. Dual-band polarization conversion metasurface for rcs reduction. IEEE Transacation on Antennas Propagation. 2020;**69**(5):3044-3049. DOI: 10.1109/TAP.2020.3028148

[57] Kamal B, Chen J, Yin Y, Ren J, Ullah S, Khan WUR. High efficiency and ultr wideband polarization converter based on L-shaped metasurface. Optical Material Express. 2021;**5**(1):1343-1352. DOI: 10.1034/OME.423324

[58] Ding F, Pors A, Bozhevolnyi SI. Gradient metasurfaces: A review of fundamentals and applications. Reports on Progress in Physics. 2018;**81**(2): 026401. DOI: 10.1088/1361-6633/aaa8732

[59] Li A, Singh S, Sievenpiper D. Metasurfaces and their applications. Nano. 2018;**7**(6):989-1011. DOI: 10.1515/nanoph-2017–0120

[60] Smith D, Schultz S, Markois P, Soukoulis C. Determination of effective permittivity and permeability of metamaterials from reflection and transmission coefficients. Physical Review B. 2002;**65**(19):195104. DOI: 10.1103/PhysRevB.65.195104

[61] Numan AB, Sharawi MS. Extraction of material parameters for metamaterials using a full-wave simulator [education column]. IEEE Antennas and Propagation Magazine. 2013;**55**(5):202-211. DOI: 10.1109/MAP.2013.6735515

[62] Khan B, Kamal B, Ullah S, Khan I, Shah JA, Chen J. Design and experimental analysis of dual-band polarization converting metasurface for microwave applications. Scientific Reports. 2020;**10**(1):1-13. DOI: 10.1038/s41598-020-71959-y

[63] Abdullah KB, Ullah S, Khan B, Rahim T, Ren J. Ultrathin and high efficient wideband polarization converting metasurface. International Journal of RF and Microwave Computer-Aided Engineering. 2021;**31**(7):e22673. DOI: 10.1002/mmce.22673

Section 2

Metamaterials in Antennas

Chapter 2

Study on Miniaturization of Antenna Using Metamaterials

Andrews Christina Josephine Malathi

Abstract

Metamaterials (MTMs) are artificially built materials intended to give its properties from the internal structure, rather than the chemical composition found in natural materials. Electric permittivity (ε) and magnetic permeability (μ) are the two basic parameters which describe the electromagnetic property of a material or medium. Permittivity describes how a material is affected when it is placed in electric field. And permeability describes how a material is affected in presence of magnetic field. Metamaterials may have either negative permittivity or permeability or both may be negative simultaneously. The concept of metamaterials has additionally been utilized to design different kinds of patches with upgraded performance, such as improved gain and enhanced efficiency. Also, it has been utilized for the scaling down of patches. Two parameters are utilized in the collected works for antennas using metamaterials. We can adjust the refractive index of the metamaterial to positive, near-zero or negative values. Utilization of epsilon negative, MNG (μ - Mu negative) or DNG (double negative) are called metamaterial- based antennas and the use of metamaterial unit cell for example complementary split ring resonator, split ring resonator and so on are alluded as metamaterial inspired antennas. The design of complementary split ring resonator and its equivalent circuit will be discussed in this work. CSRR (complementary split ring resonator) provides both isolation enhancement and miniaturization for MIMO antenna.

Keywords: MTM (metamaterials), CSRR (complementary split ring resonator), left handed materials, NRW (Nicolson Ross weir), MNG (Mu negative), DNG (Double negative)

1. Introduction

Jagadish in 1898 developed the first metamaterial out of artificial chiral components [1]. Lindman, in 1914 made artificial chiral media by installing numerous randomly arranged little wire helices in the host medium [2]. Kock made microwave lenses by inserting metallic strips, wires and disks occasionally so as to tailor the artificial media refractive index [3]. The historical backdrop of left- handed materials can be explained by Veselago's hypothetical speculation in 1968 [4], in which he showed that the LHMs could give rise to a surprising optical phenomenon as light passes through them. The UCSD (University of California, San Diego) group in 2000, which followed the work by Pendry et al. [5–7] synthesized the principal left-handed material [8, 9]. As manufacture strategies enhance, the incorporations (artificial structures) inserted in the host medium decrease in size [9].

Electromagnetic waves interact with metamaterials in the host medium, induce electric and magnetic moments, which thus influence the material's transmission abilities and material parameters, for example, permeability and permittivity. While portraying the permittivity (ε) and permeability (μ) of a metamaterial, one must describe the structure as homogeneous. If one considers a solitary incorporation as a major aspect of a unit cell periodically installed inside the host medium, its size p must be not exactly a fourth of the incident radiation wavelength: $p < \lambda/4$ [9]. This homogeneity relation is a standard guideline condition. The relation is normally utilized in recognizing lumped parts from quasi-lumped components, ($\lambda/4 < p < \lambda/2$) and distributed components: $p > \lambda/4$ [9]. This condition guarantees that refractive phenomena inborn in homogeneous materials command over scattering and diffusion effects. The electromagnetic radiation is basically uninformed of the lattice structure in the host medium and keeps up field uniformity in the direction of propagation inside the structure.

Metamaterials are designed to make materials with almost zero estimations of permittivity; negative permeability or permittivity. A material with just negative (ε) is called ENG and only negative (μ) is assigned to MNG material. Permittivity describes how a material is affected when it is placed in electric field. And permeability describes how a material is affected in presence of magnetic field. Material with both negative permeability and permittivity is called double negative (DNG). Over the previous decade, various arrangements of metamaterials have been approved which when organized intermittently, showed metamaterial properties for a specific range of frequencies [9]. This left handed material made utilization of an array of conducting, nonmagnetic components, continuous wires like circular shapes etc. to accomplish a negative effective permittivity and negative effective permeability.

Despite the fact that optical characteristics are completely determined by relative parameters ε and μ, the refractive index (n) is frequently utilized. n might be found from Eq. (1) [10],

$$n = \pm\sqrt{\varepsilon\mu} \tag{1}$$

Practically all see-through materials, for example, water or glass, has positive values for both ε and μ [10]. Numerous metals, (for example, gold and silver) has negative ε at observable wavelength [10]. Materials having one or the other ε or μ negative is impervious to EM emission. For the above - mentioned materials, a positive square root is utilized for n by tradition. In any case, some materials having $\varepsilon < 0$ and $\mu < 0$; since their product is positive, n is additionally real. In such conditions, negative square root of n is taken. The prior considerations are simple for materials, having complex values of ε and μ. The real parts of both do not need to be negative [10].

Electromagnetics scientists regularly utilize the term LHM, barely, for the materials having negative refractive index. A contrast of refraction in a LHM to a typical material is explained in **Figure 1** [11]. The Quadrant 1, of **Figure 2** the Electric, magnetic (H) field and the wave 'k' structure a right handed system, as given by Maxwell's equations. The second quadrant of **Figure 2** ($\varepsilon < 0$ and $\mu > 0$) shows electric plasmas which bolster evanescent waves. The fourth quadrant ($\varepsilon > 0$ and $\mu < 0$) additionally bolsters evanescent waves. These are both single negative (SNG)

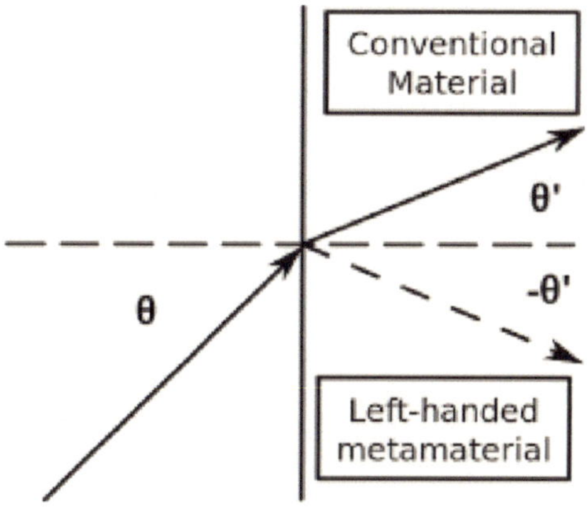

Figure 1.
Refraction in a left-handed metamaterial.

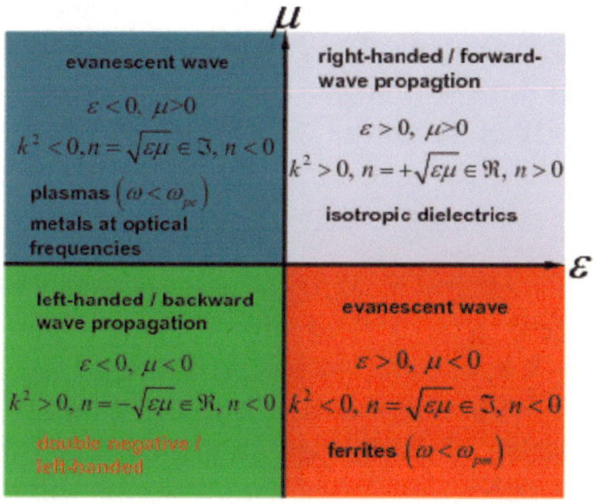

Figure 2.
Comparison of material with different permittivity and permeability, courtesy—http://www.ihf.uni-stuttgart.de.

quadrants, implying that just a single parameter is negative. The third quadrant ($\varepsilon < 0$ and $\mu < 0$ or double negative (DNG)) contains the left handed materials, which were proposed in 1967 [11]. In LHM, the E, H field and the wave vector k structure a left handed system and these help "in reverse" proliferating waves i.e. negative refraction occurs such that light and other radiation gets bent backwards as it enters the structure [11]. The term reverse refers to the opposite sign of group and phase velocity. The index of refraction $n = \pm\sqrt{\varepsilon_r \mu_r}$ is negative. A comparison of the materials is shown in **Figure 2**.

The light beams would be refracted on a similar side perpendicular when coming inside the material, since n_2 is negative. A light source advancing on the way to an eyewitness seems to reduce its frequency, as the doppler shift is reversed. Cherenkov focuses in another way about the radiation delivered by a quick moving molecule as it goes through a media [11]. Phase velocity is antiparallel to the time averaged poynting vector of wave 'k'. This implies that unlike RHM, the wave fronts are heading backwards to the movement of energy [12]. Plane wave propagation spreading in MTMs, the E, H field and wave propagation abide by a left hand rule, and hence named as left-handed metamaterials [13]. The impact of negative refraction is undifferentiated from wave transmission in a left-handed transmitting line.

2. Applications of metamaterials

MTMs are of main importance in electromagnetics [14]. All together for its physical shape to influence EM waves, a MTM should have physical features small than λ of the EM radiation it connects with. For example, if a MTM is to carry on as a homogenized material precisely depicted by refraction index, the element dimensions need to be a lot littler than the λ. For the visible light having wavelengths of less than one μm commonly, the structures are usually half or not larger than 280 nm in dimensions [14]. The structures need to be of one decimeter [14] for microwave radiation. They are quite often artificial, developed as periodic arrangement of current conducting components, (for example, circles of wire) which have appropriate inductive and capacitive attributes [15].

2.1 Superlens

The materials having negative refractive index, enables to focus near field light (the near-field (or **evanescent**) light consists of a non-propagating field that exists near the surface of an object at distances less than a single wavelength of light) therefore making an ideal super lens. The first super lens gave resolution three times better than the diffraction limit which was shown at microwave frequencies (1–1000 GHz) at University of Toronto [16].

2.2 Cloaking devices

MTMs are used for constructing cloaking devices as well. These components usually include surrounding the item to be cloaked with a casing which influences the entry of light close to it [17]. A US-British group of researchers made a MTM that made an article undetectable to microwave emission in october 2006 [17]. Since light is simply one more type of EM radiation, this was viewed as the initial move to a cloaking gadget for visible light, however further developed nano systems will be required because of noticeable light small wavelengths. Two engineers from Purdue university declared a hypothetical plan for an optical cloaking gadget dependent on the British idea on April 2, 2007 as explained below. The plan sends a periodic arrangement of minor needles sticking out by a centre rod which will create an object inside the cloak undetectable in a λ of 632.8 nm [18]. Duke University and Imperial College London are as of now looking into this utilization of MTMs and has figured out how to cloak an item in microwave range utilizing exceptional concentrical rings [18]. The microwave frequencies were hardly influenced by the nearness of the cloaked object.

2.3 Agile antennas

Metamaterials are additionally designed for structuring agile antennas [19]. Investigation at National Institute of Standards and Technology in US has shown that slender films created of MTMs can enormously diminish dimensions of resonant circuits which create microwaves, possibly empowering much small mobile phones and other microwave gadgets [19].

3. Design of metamaterials (complementary split—ring resonator)

By the parametric investigations and examination of the patch, a design methodology for miniaturizing antenna for variable frequency range utilizing the CSRR technique was created [20–24]. A CSRR is counter duplicate of split ring resonator, which is made by eliminating the conductor from the outline of SRR within the ground structure. It is a resonant structure utilized broadly in the investigation of MTMs [20]. With recurrent placement of the CSRR, negative ε permittivity can be obtained near its resonant frequency. Its frequency is equivalent to SRR of a similar size and it is demonstrated as a LC circuit [25]. The equivalent LC circuit analysis is discussed in Section 3.5. Distinctive forms of SRR and CSRR are shown to exist in different works, however the dual slit rounded and rectangular split ring resonator are usually considered and utilized due to the ease of creating their structures [25]. The CSRR consist of two concentrical or rectangular rings with an opening in each ring. The procedure is described below in **Figure 3**.

1. Design the patch antenna utilizing the transmission line model for a frequency double as that of the needed (5 GHz). 2. The patch antenna is simulated using high frequency structure simulator software (HFSS). 3. A CSRR is engraved beneath the patch in such a way that the outward radius of the CSRR is encasing the boundaries of the antenna. 4. Simulate and locate the frequency of the patch. 5. If it is smaller than anticipated, adjust by decreasing the radius 'r' or by changing spacing 'k' and/or width 'w'. 6. During fine tuning, if the frequency is beyond as expected, move the feed alongside the antenna element till the resonant frequency is achieved. 7. Amid this procedure, ensure the patch to remain as primary radiating element.

3.1 Design of antenna and its geometry

A circular patch antenna with dimensions 25 × 25 mm^2 was at first intended to resonate at 5 GHz utilizing FR4 substrate with a dielectric constant of 4.4 and

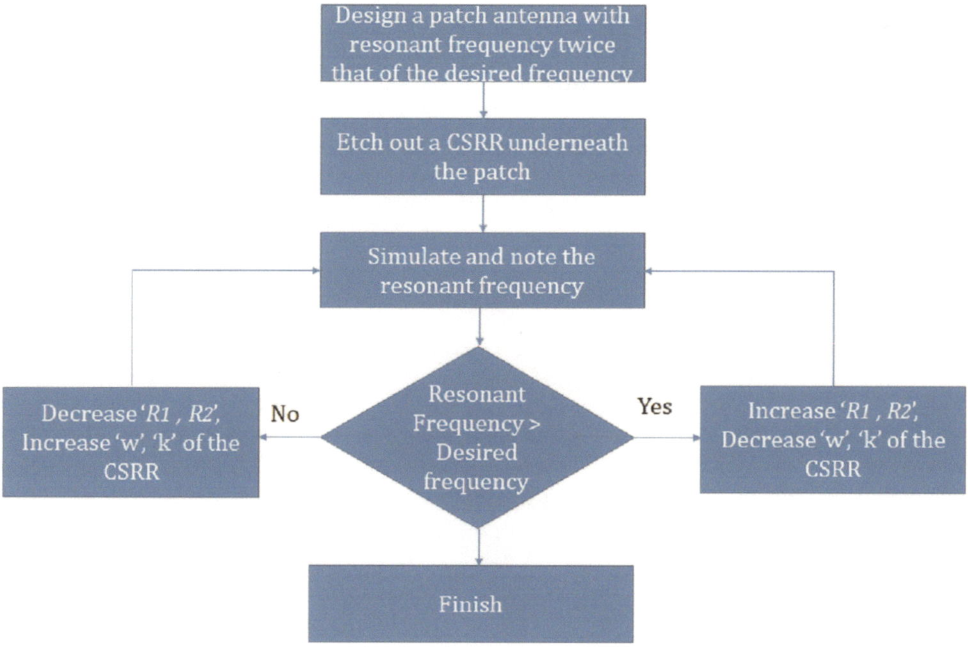

Figure 3.
A flow chart representation for the design of antenna miniaturization using CSRR.

thickness 1.6 mm [26]. The patch antenna was excited utilizing 50 Ω microstrip line. The geometry of the antenna is shown in **Figure 4**.

3.2 Parametric analysis of the antenna design

Initially the antenna was designed at a frequency (5 GHz) higher than the desired resonant frequency (2.34 GHz). The radius obtained was 8.39 mm with feed length of 8.25 mm and feed width was 3.05 mm. A parametric analysis was carried out for various radius of the circular patch starting from 7 mm and the optimal radius of 8.39 mm is chosen as shown in **Figure 5**. It was observed that when the patch radius increases the resonant frequency decreases.

Similarly the analyses were carried out for different length of the feed with fixed radius of 8.39 mm and it was seen that as the feed length increases the return loss increases providing better impedance matching as seen in **Figure 6**. A feed length of $\frac{\lambda_g}{2}$ was chosen with 8.25 mm.

3.3 Antenna geometry with CSRR

Methodology given in the flow chart, **Figure 3**, was followed. Parametric analysis is performed using HFSS to determine the dimension of CSRR given in **Figure 7**. After simulation, if the obtained frequency is greater than the resonant frequency, the CSRR dimensions were altered by expanding the radius of CSRR and diminishing the width of the ring and spacing between the rings. If the frequency is smaller than the resonant frequency, at that point the CSRR is altered by diminishing the radius of CSRR and expanding the width and spacing between the rings. The dimensions of the CSRR were changed to adjust at 2.34 GHz. When the radius of the ring decreased with increase in the width and spacing between the rings, the frequency increased.

The geometry of the antenna with CSRR at the ground plane is shown in **Figure 8** and its dimensions are given in **Figure 9**. The radius of the inner ring of the CSRR is R_1, radius of the outer ring of the CSRR is R_2, the width of each ring is 'w', the spacing between the rings is 'g' and the width of slit in the rings is 'k'. After including the

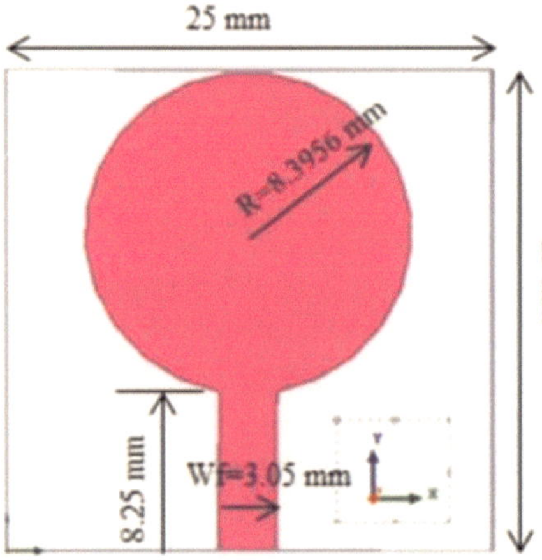

Figure 4.
Single patch antenna structure.

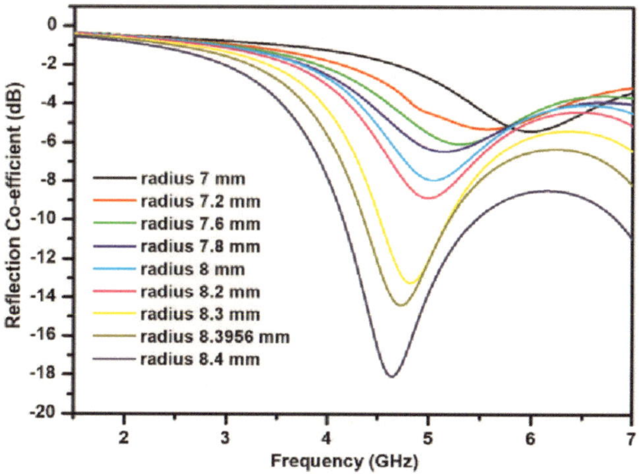

Figure 5.
Reflection co-efficient of the antenna vs. frequency for different patch radius.

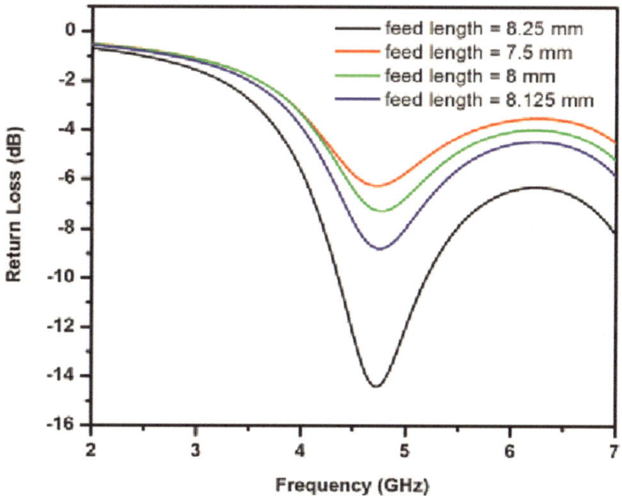

Figure 6.
Reflection co-efficient of the for various feed length CSRR.

CSRR, the radius of the patch was optimized to 7.65 mm providing 2.34 GHz resonant frequency.

3.4 Parametric analysis for the position of CSRR

To study the effect of CSRR loading on the patch, parametric study is performed by varying the position of the CSRR. As noted in **Figure 10** when the CSRR is shifted away from the feed the resonating frequency is increasing and when it is near to the feed i.e. (c = 0 mm), it provided an impedance matching at 2.34 GHz with return loss of −17 dB.

3.5 Analysis of permittivity for complementary split - ring resonator (CSRR)

Metamaterials (MTM) are utilized for isolation improvement between nearby components because of the presence of a band gap in their frequency response [27].

S.No	Outer Radius R_2 (mm)	Inner Radius R_1 (mm)	Width of the rings (*w*)	Spacing between the rings (*k*)	Impedance bandwidth (GHz)	Return Loss (dB)	% increase or reduction in frequency
1	4.1938	3.4838	0.5	0.21	2.47 to 2.51	-12	7.3% increase
2	**4.1938**	**3.4938**	**0.5**	**0.2**	**2.3 to 2.37**	**-17**	**-**
3	4.2	3.5369	0.48	0.1831	2.27 to 2.32	-12	1.3% decrease
4	4.2	3.537	0.49	0.183	2.26 to 2.32	-13.5	1.73% decrease

Figure 7.
Parametric analysis for design of antenna vs. frequency.

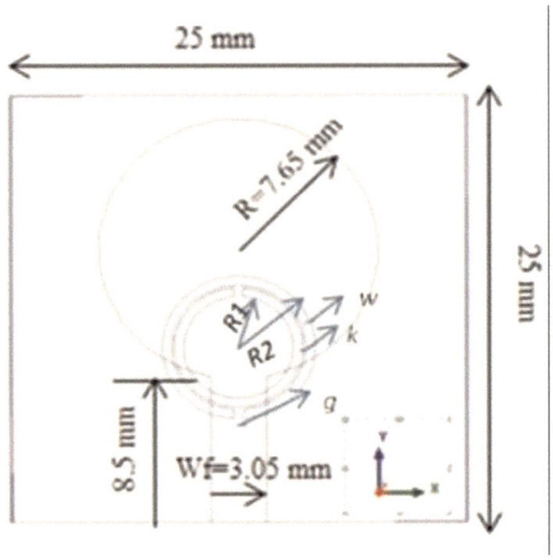

Figure 8.
Single patch antenna with CSRR structure.

Description	Parameter	Dimensions
Inner radius of the CSRR ring	R_1	3.4938
Outer radius of the CSRR ring	R_2	4.1938
Width of the rings	w	0.5 mm
Spacing between the rings	g	0.5 mm
Width of the slit	k	0.2 mm

Figure 9.
Dimensions of CSRR.

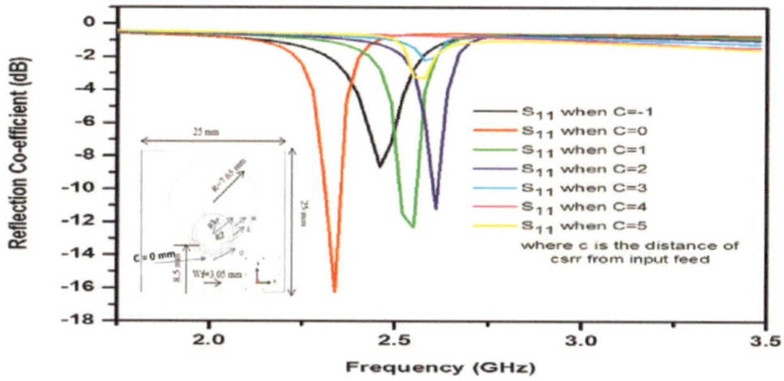

Figure 10.
CSRR position variation.

A significant decrease in patch antenna size was additionally seen by CSRR stacking on a patch antenna [28]. The two most generally utilized metamaterial structures for isolation enhancement between neighboring components are the utilization of split ring resonators (SRR) and complementary split ring resonator (CSRR) [27].

The resonant frequency of a SRR is equivalent to that of a CSRR of a similar measurement. CSRRs are the opposite image of SRRs (Babinet's standard), and an axial time changing E field is important to energize the rings that make a feasible negative ε medium and hinder signal transmission at resonance. CSRRs are used as a negative permittivity bandstop filter. Band stop filter blocks and rejects frequencies that lie between its two cut-off frequency point's passes all those frequencies either side of this range. CSRR is a resonant structure which acts as an LC circuit [29]. It interacts with the axial E field and displays negative permittivity close to the resonant frequency. The unit cell of CSRR is shown in **Figure 11**. Antenna miniaturization by loading CSRR is studied through unit cell analysis and equivalent circuit analysis. Unit cell CSRR was simulated with HFSS.

Relative permittivity is to comprehend the behavioral attributes of CSRR. The equations given by Nicolson, Ross and Weir (NRW) empower the calculation of the complex permittivity and permeability of a material from the measured S-parameters. S-parameters describe the input-output relationship between ports (or terminals) in an electrical system. NRW had shown the extraction of relative permittivity utilizing S parameters [25, 30–32] and is given by,

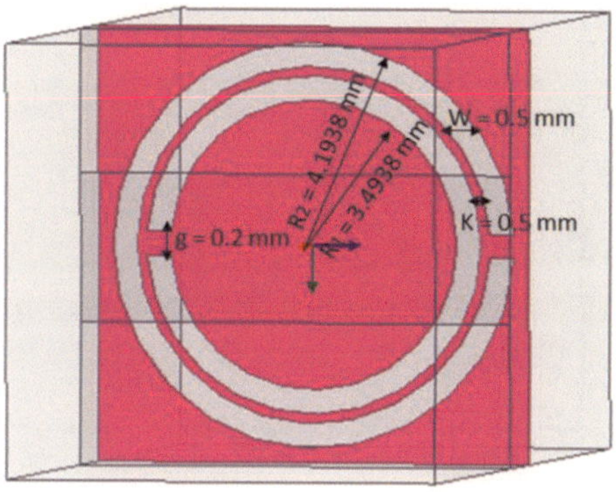

Figure 11.
Unit cell complementary Split - ring resonator.

$$V_1 = S_{11} + S_{21} \tag{2}$$

$$V_2 = S_{21} - S_{11} \tag{3}$$

$$\varepsilon_r = \frac{2c(1 - V_1)}{\omega di(1 + V_1)} \tag{4}$$

$$\mu_r = \frac{2c(1 - V_2)}{\omega di(1 + V_2)} \tag{5}$$

Where ω is the frequency in radian, c is the velocity of light in m/s, d is the thickness of the substrate, i correspond to imaginary part and V_1 is the voltage maxima in volts and V_2 is the voltage minima in volts. From the S parameters, relative permittivity was plotted with Eqs. (2)–(5). The negative estimation of relative permittivity was acquired at the ideal resonant frequency of 2.34 GHz plainly displaying that CSRR goes about as a band stop filter as in **Figure 12**.

3.6 Equivalent circuit analysis for CSRR

The equivalent circuit analysis for the unit cell CSRR for the proposed work is discussed. The lumped element equivalent model of the unit cell CSRR is modeled in AWR microwave office software as shown in **Figure 13** [32] and the simulation results of unit cell in HFSS mentioned in **Figure 11** are compared and its results are given in **Figure 14**. Both the results are in good agreement. The transmission line is represented by the MLIN in AWR software, parallel LC circuit (C_2, L_3) represents the resonant circuit, the current flowing through the rings representing the inductance (L_1, L_2) and capacitor (C_1) indicates the electric coupling between the transmission line and CSRR (**Figure 13**).

The unit cell CSRR was analyzed using NRW method to check for the negative permittivity of CSRR. Its lumped element equivalent circuit model of the CSRR was analyzed in AWR and verified with the HFSS results.

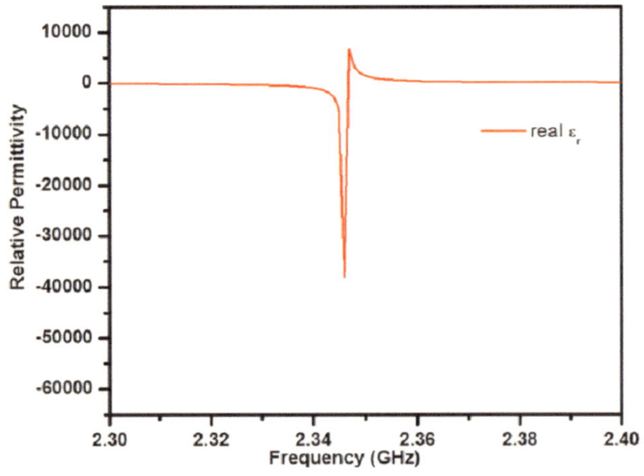

Figure 12.
Relative permittivity vs. frequency for a unit cell CSRR for LTE band 40.

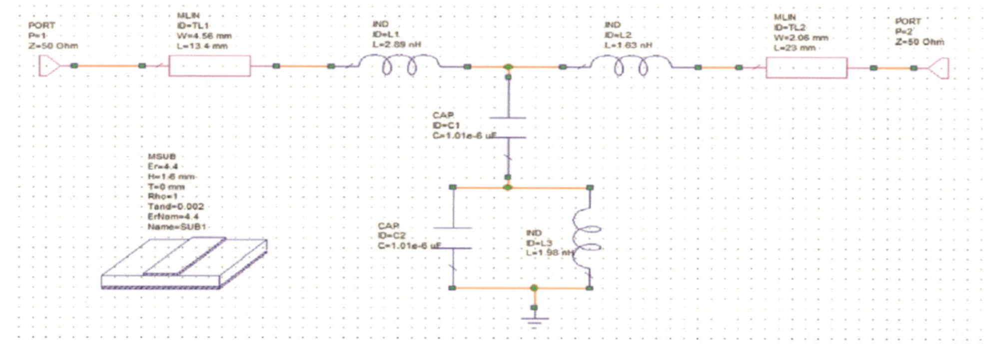

Figure 13.
Equivalent circuit for a unit cell CSRR for band 40.

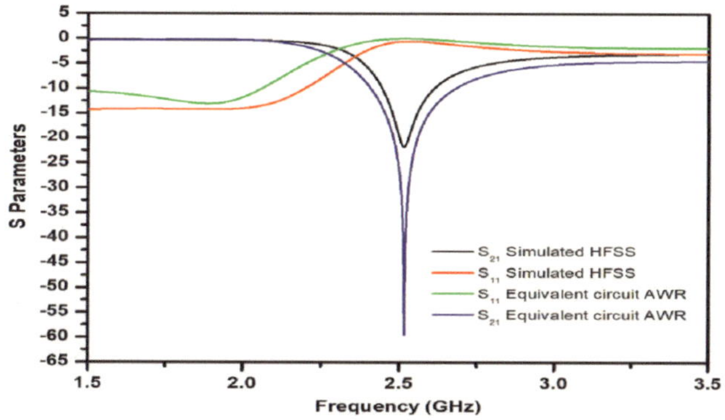

Figure 14.
S parameter for the equivalent unit cell CSRR circuit for band 40.

4. Antenna miniaturization

To minimize the dimensions of the antenna, a CSRR was etched out beneath the patch in the ground plane. The antenna with CSRR resonates at 2.34 GHz and without CSRR resonates at 4.8 GHz as shown in **Figure 15** covering a band of 2.3 to 2.374 GHz occupying the LTE band long term evolution (2.3-2.4) GHz. If the antenna is designed for 2.34 GHz, the radius of the patch would be 17 mm. But by introducing CSRR, the antenna designed for 5 GHz has radius 7.621 mm with a shift in the resonant frequency to 2.34 GHz providing a size reduction of 55.17%.

4.1 MIMO antenna geometry

The simulation was extended for MIMO configuration by designing two circular patch antennae with dimensions ($50 \times 25 \times 1.6$ mm^3) shown in **Figure 16**.

The front and back view of the photograph of the fabricated antenna is shown in **Figure 17a** and **b**, respectively.

A stub was introduced between the two patches. The parametric analysis for the width and length of the stub was carried out using HFSS as shown in **Table 1**. Without stub the S_{11} parameter is 2.30 to 2.37 and S_{22} is 2.33 to 2.40 GHz and with the stub the S_{11} parameter is 2.30 to 2.37 and S_{22} is 2.31 to 2.38 GHz.

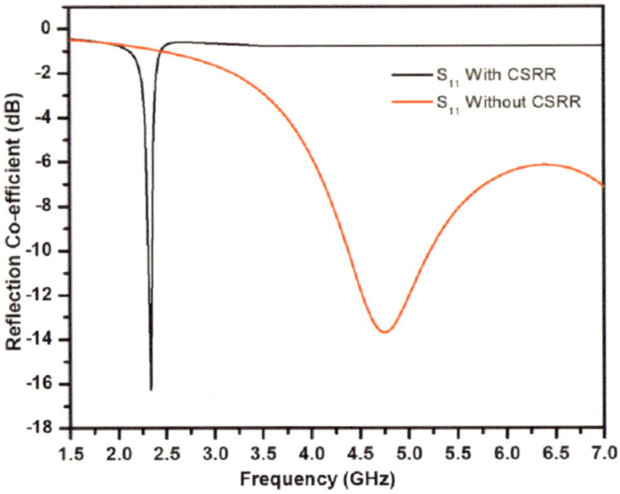

Figure 15.
Comparison of reflection co-efficient with and without CSRR for single patch antenna.

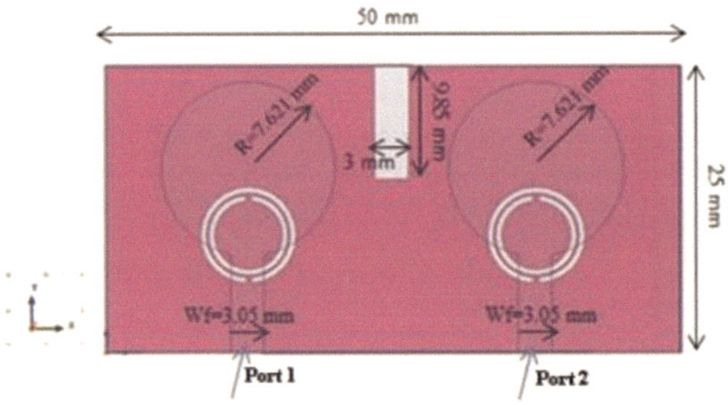

Figure 16.
2 × 1 MIMO antenna structure.

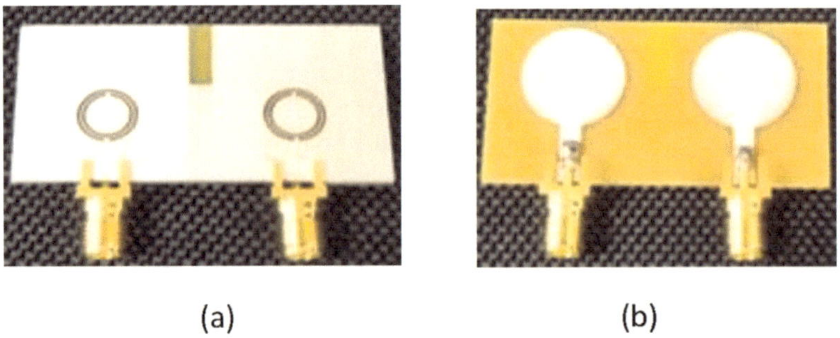

Figure 17.
Fabricated antenna (a) front view (b) back view.

With the introduction of the stub, there is a shift in frequency for S_{22}. When the length of the stub is decreased, keeping width constant S_{22} is shifted towards left (lower frequency range - i.e. frequency reduced with increased physical length of the

S. no	Stub length (l_s) (mm)	Slub width (w_s) (mm)	S_{11} (Imp BW)	S_{22} (Imp BW)	Inference
1	17	3	2.36–2.43	2.38–2.45	Both S_{11} and S_{22} are shifted beyond 2.3 to 2.4 GHz range
2	17	2	2.3–2.38	2.34–2.47	S_{11} shifted from 2.30 to 2.31 GHz and S_{22} shifted from 2.33 to 2.34 GHz
3	15	3	2.40–2.48	2.40–2.48	Both S_{11} and S_{22} are shifted to high frequency range beyond 2.40 GHz
4	12	2	2.28–2.35	2.34–2.41	S_{11} shifted from 2.30 to 2.28 GHz and S_{22} shifted from 2.33 to 2.34 GHz
5	**9.8**	**3**	**2.30–2.371**	**2.31–2.388**	**S_{22} shifted from 2.33 to 2.31 GHz low frequency range**

Table 1.
Parametric analysis of stub length and width.

stub). When the width of the stub is decreased, S_{11} is shifted towards left (lower frequency range).

4.2 Return loss, bandwidth, and isolation characteristics

The simulation result for the return loss and isolation were observed to be −30 dB and − 35.51 dB at the resonant frequency of 2.34 GHz and the measured results for the return loss and isolation were − 27 dB and − 33.5 dB, respectively, at the resonant frequency of 2.3 GHz as shown in **Figures 18** and **19**. The designed antenna covered a bandwidth of 2.3–2.372 GHz using simulation whereas 2.26–2.34 GHz for the measurement.

It was observed that there was a shift in resonating frequency of second patch and hence a stub was inserted between two patches. The purpose of stub between the two

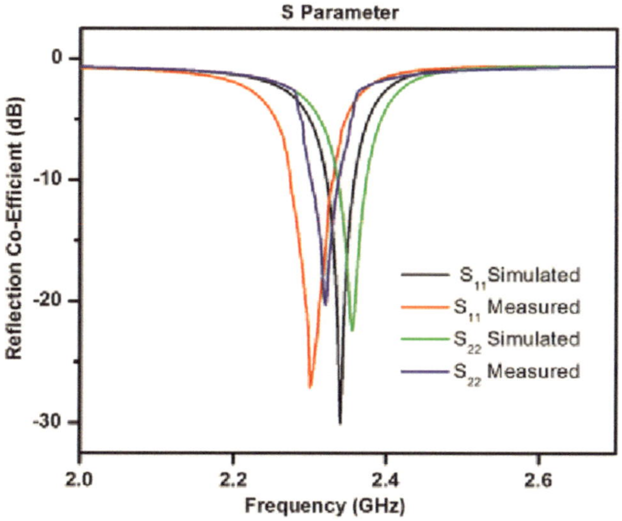

Figure 18.
Comparison of measured and simulated reflection co-efficient characteristics for MIMO antenna.

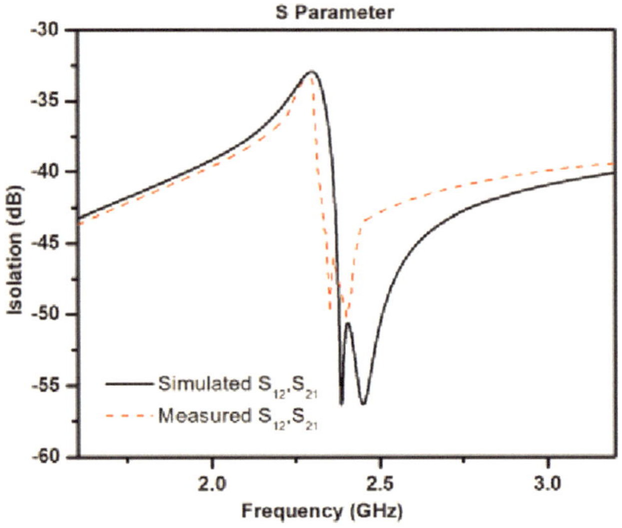

Figure 19.
Comparison of measured and simulated isolation characteristics for MIMO antenna.

antenna elements is in increasing the electrical length between the two elements thereby there is a shift in the resonant frequency of second patch in the lower side i.e. (2.354 GHz). S_{11} occurs at 2.34 GHz and S_{22} occurs at 2.367 GHz. By introducing the stub between the patches, there is an increase in the current path which shifts the frequency for S_{22} from 2.367 to 2.354 GHz. Also, the isolation between the two elements is increased from −33 dB to −35 dB with the introduction of the stub as shown in **Figure 20**.

The simulated results for the return loss without stub was observed to be −27.35 dB and with stub is −30 dB at the resonant frequency of 2.34 GHz with an isolation of −33.3 dB without stub covering a bandwidth of 2.3–2.373 GHz and the comparative plot for both with and without stub is shown in **Figure 20**.

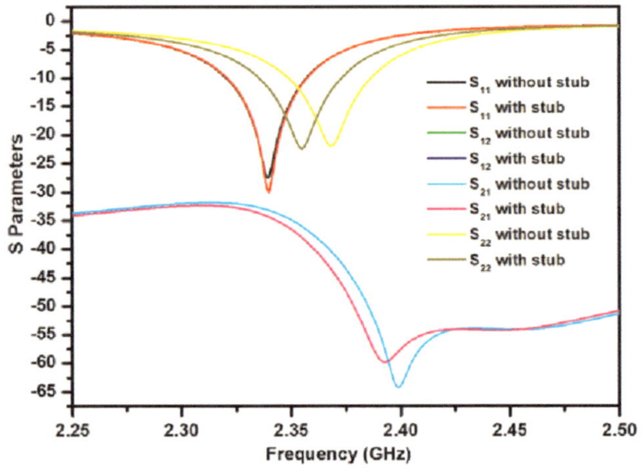

Figure 20.
Comparison of reflection co-efficient and isolation for MIMO antenna with stub and without stub.

5. Conclusions

Using CSRR, a size reduction of 55.17% was achieved. The antenna designed covered a bandwidth of 2.3–2.374 GHz with a maximum return loss of −27 dB at 2.34 GHz and isolation of −33.5 dB amid the ports. The antenna performance is suitable for LTE band 40 operation. Here the analysis of CSRR with respect to the permittivity was discussed and its equivalent circuit was analyzed. Also the antenna miniaturization was discussed by incorporating the CSRR. The antenna resonated at a frequency of 4.8 GHz. With the introduction of CSRR beneath the patch in the ground plane the antenna resonated at a frequency of 2.34 GHz providing a size reduction of 55.17%. If the antenna was designed for 2.34 GHz, the dimension of the patch would be 17 mm but by introducing CSRR, the antenna designed for 5 GHz of dimension 7.621 mm there is a shift in the resonant frequency to 2.34 GHz providing a size reduction of 55.17%.

Author details

Andrews Christina Josephine Malathi
Vellore Institute of Technology (VIT), Vellore, India

*Address all correspondence to: achristina@vit.ac.in

References

[1] Bose JC. On the rotation of plane of polarisation of electric wave by a twisted structure. Proceedings of the Royal Society of London. 1898;**63**(389-400):146-152

[2] Lindell IV, Sihvola AH, Kurkijarvi J, Lindman KF. The last Hertzian, and a harbinger of electromagnetic chirality. IEEE Antennas and Propagation Magazine. 1992;**34**(3):24-30

[3] Kock WE. Metallic delay lenses. Bell System Technical Journal. 1948;**27**(1):58-82

[4] Veselago VG. The electrodynamics of substances with simultaneously negative values of ε and μ. Physics-Uspekhi. 1968; **10**(4):509-514

[5] Pendry JB, Holden AJ, Stewart WJ, Youngs I. Extremely low frequency plasmons in metallic mesostructures. Physical Review Letters. 1996;**76**(25):4773

[6] Pendry JB, Holden AJ, Robbins DJ, Stewart WJ. Magnetism from conductors and enhanced nonlinear phenomena. IEEE Transactions on Microwave Theory and Techniques. 1999;**47**(11):2075-2084

[7] Pendry JB. Negative refraction makes a perfect lens. Physical Review Letters. 2000;**85**(18):3966

[8] Smith DR, Padilla WJ, Vier DC, Nemat-Nasser SC, Schultz S. Composite medium with simultaneously negative permeability and permittivity. Physical Review Letters. 2000;**84**(18):4184

[9] Shelby RA, Smith DR, Schultz S. Experimental verification of a negative index of refraction. Science. 2001; **292**(5514):77-79

[10] Steinmetz. The Natural Period of a Transmission Line and the Frequency of Lightning Discharge Therefrom. The Electrical World. 1898. pp. 203-205. [Thesis]

[11] Lee D, Nguyen DM, Rho J. Acoustic wave science realized by metamaterials. Nano Convergence. 2017;**4**(1):1-5

[12] Liu Q, Hall PS, Borja AL. Efficiency of electrically small dipole antennas loaded with left-handed transmission lines. IEEE Transactions on Antennas and Propagation. 2009;**57**(10): 3009-3017

[13] Herraiz-Martinez FJ, Hall PS, Liu Q, Segovia-Vargas D. Tunable left-handed monopole and loop antennas. In: 2009 IEEE Antennas and Propagation Society International Symposium. Charleston, SC: IEEE; 2009. pp. 1-4

[14] Caloz C, Itoh T. Application of the transmission line theory of left-handed (LH) materials to the realization of a microstrip" LH line". In: IEEE Antennas and Propagation Society International Symposium (IEEE Cat. No. 02CH37313). Vol. 2. San Antonio, TX, USA: IEEE; 2002. pp. 412-415

[15] Sihvola A. Metamaterials in electromagnetics. Meta. 2007;**1**(1):2-11

[16] Grbic A, Eleftheriades GV. Overcoming the diffraction limit with a planar left-handed transmission-line lens. Physical Review Letters. 2004; **92**(11):117403

[17] Cai W, Chettiar UK, Kildishev AV, Shalaev VM. Optical cloaking with metamaterials. Nature Photonics. 2007; **1**(4):224-227

[18] Hao J, Yan W, Qiu M. Super-reflection and cloaking based on zero index metamaterial. Applied Physics Letters. 2010;**96**(10):101109

[19] Boutayeb H, Denidni TA, Mahdjoubi K, Tarot AC, Sebak AR, Talbi L. Analysis and design of a cylindrical EBG-based directive antenna. IEEE Transactions on Antennas and Propagation. 2006;**54**(1):211-219

[20] Dong Y, Toyao H, Itoh T. Design and characterization of miniaturized patch antenna loaded with complementary split ring resonators. IEEE Transactions on Antennas and Propagation. 2012; **60**(2):772-785

[21] Ouedraogo RO, Rothwell EJ, Diaz AR, Fuchi K, Temme A. Miniaturization of patch antennas using a metamaterial-inspired technique. IEEE Transactions on Antennas and Propagation. 2012;**60**(5):2175-2182

[22] Khan MU, Sharawi MS. A compact 8-element MIMO antenna system for 802.11 ac WLAN applications. In: 2013 International Workshop on Antenna Technology (iWAT). IEEE; March 2013. pp. 91-94

[23] Habashi A, Nourinia J, Ghobadi C. Mutual coupling reduction between very closely spaced patch antennas using low-profile folded split-ring resonators (FSRRs). IEEE Antennas and Wireless Propagation Letters, IEEE. 2011;**10**: 862-865

[24] Saraswat RK, Kumar M. Miniaturized slotted ground UWB antenna loaded with metamaterial for WLAN and WiMAX applications. Progress in Electromagnetics Research B. 2016;**65**:65-80

[25] Naqui J, Durán-Sindreu M, Martín F. Modeling split-ring resonator (SRR) and complementary split-ring resonator (CSRR) loaded transmission lines exhibiting cross-polarization effects. IEEE Antennas and Wireless Propagation Letters. 2013;**12**:178-181

[26] Malathi AC, Thiripurasundari D. Compact 2× 1 MIMO antenna system for LTE band. Progress in Electromagnetics Research C. 2017;**75**:63-73

[27] Baena JD, Bonache J, Martín F, Sillero RM, Falcone F, Lopetegi T, et al. Equivalent-circuit models for split-ring resonators and complementary split-ring resonators coupled to planar transmission lines. IEEE Transactions on Microwave Theory and Techniques. 2005;**53**(4):1451-1461

[28] Dong Y, Toyao H, Itoh T. Design and characterization of miniaturized patch antennas loaded with complementary split-ring resonators. IEEE Transactions on Antennas and Propagation. 2011; **60**(2):772-785

[29] Cheng X, Senior DE, Kim C, Yoon YK. A compact omnidirectional self-packaged patch antenna with complementary split-ring resonator loading for wireless endoscope applications. IEEE Antennas and Wireless Propagation Letters. 2011;**10**: 1532-1535

[30] Ziolkowski RW. Design, fabrication, and testing of double negative metamaterials. IEEE Transactions on Antennas and Propagation. 2003;**51**(7): 1516-1529

[31] Rothwell EJ, Frasch JL, Ellison SM, Chahal P, Ouedraogo RO. Analysis of the Nicolson-Ross-Weir method for characterizing the electromagnetic properties of engineered materials. Progress in Electromagnetics Research. 2016;**157**:31-47

[32] Wirgin A. Retrieval of the frequency-dependent effective permeability and permittivity of the inhomogeneous material in a layer. Progress in Electromagnetics Research B. 2016;**70**:131-147

Chapter 3

Manipulating Light with Tunable Nanoantennas and Metasurfaces

Davide Rocco, Andrea Locatelli, Domenico De Ceglia, Andrea Tognazzi, Attilio Zilli, Michele Celebrano, Marco Finazzi, Antonio Ferraro, Roberto Caputo and Costantino De Angelis

Abstract

The extensive progress in nanofabrication techniques enabled innovative methods for molding light at the nanoscale. Subwavelength structured optical elements and, in general, metasurfaces and metamaterials achieved promising results in several research areas, such as holography, microscopy, sensing and nonlinear optics. Still, a demanding challenge is represented by the development of innovative devices with reconfigurable optical properties. Here, we review recent achievements in the field of tunable metasurface. After a brief general introduction about metasurfaces, we will discuss two different mechanisms to implement tunable properties of optical elements at the nanoscale. In particular, we will first focus on phase-transition materials, such as vanadium dioxide, to tune and control the resonances of dipole nanoantennas in the near-infrared region. Finally, we will present a platform based on an AlGaAs metasurface embedded in a liquid crystal matrix that allows the modulation of the generated second harmonic signal.

Keywords: tunability, metasurface, sensor, nonlinearity, dielectric

1. Introduction

Ultra-flat optical devices allow to manipulate and control propagating light by governing its phase, amplitude, and polarization states [1, 2]. Such properties, when optimized, can be employed to perform optical operations, such as beam steering and wavefront shaping.

This can find application in several fields such as microscopy, holography, imaging, communication, and sensing [3–6]. In traditional bulk optics, the signal modifications are obtained through refraction, reflection, absorption or diffraction of light [7–9]. The working principle is based on the wave propagation through various media with different refractive indices, which are optimized to obtain a well-defined optical path for specific operating wavelengths.

This mechanism was already known to ancient populations who used, for example, primitive lenses to focus the sunlight in tiny spots in order to ignite a fire. From then on, several studies have led to an efficient optical control in several devices based on reflection and refraction, such as lenses, waveplates, and optical modulators [10–12]. Although a lot of them are still present in many commercial and technical products, their bulk nature is the main limiting factor for their integration and miniaturization in modern optical systems. In this context, a solution is offered by metasurfaces, which are artificial media composed by ensembles of subwavelength resonators (typically referred to as meta-atoms) that gained increasing attention in the last years [13–19]. Depending on the material composition, one can distinguish between metallic (or plasmonic) and dielectric metasurfaces. Both plasmonic and dielectric metasurfaces are based on the coupling between the meta-atoms to the incident electromagnetic wave. By properly designing the meta-atom shape, dimension and distribution, it is possible to implement functionalities that, in a much smaller volume, equal and even outmatch those offered by traditional bulky components, thus leading to a dramatic reduction of the footprint of the final photonic devices [20].

More specifically, devices such as metalenses, gradient phase shifters, vortex generation and hologram devices have been proposed and realized [21–26]. More recently, the focus shifted to the realization of dynamically reconfigurable metasurface [27, 28]. Differently from static metasurfaces, whose performances are defined and fixed by geometric constrains, in tunable metasurfaces the electromagnetic behavior can be modulated in response to an external stimulus, such as an electrical or optical signal. Several implementations have been proposed including ones exploiting thermal diffusion processes, optical tuning via nonlinear effects, and electron-induced phenomena [29–34]. A complete survey of these recent results is reported in [35].

Here, we review recent progresses for obtaining tunable functionalities in metasurfaces by concentrating on the material chosen as the constituent of the nanoresonators or surrounding them. The chapter is organized as follows: in Section 2 we introduce the concept of Phase-Change Materials (PCM). The PCM tuning mechanism relies on the modification of the refractive index across the phase transition. In particular, we will focus on vanadium dioxide, VO_2, and its transition from a dielectric to a metallic phase. We will demonstrate the tuning of the optical response of a plasmonic nano-resonator in the linear regime. Further, Section 3 is devoted to the reconfigurable mechanism in the nonlinear regime. More specifically, we report the tunable control of the Second Harmonic (SH) generated by an Aluminum Gallium Arsenide (AlGaAs) metasurface embedded in a liquid crystal matrix. Finally, in Section 4, we provide final remarks and an outlook on possible research directions.

2. Phase change material coupled with a plasmonic nanoantenna

In this section we report recent results concerning the use of the insulator-to-metal transition in a PCM to dynamically control the optical behavior of a nano-resonator [36, 37]. Let us underline that, in this paragraph, we will concentrate on a single nano-antennas. However, the reported results are fundamental for approaching the metasurface devices. In particular, we focus on VO_2 which is an extremely promising PCM for applications, since its insulator–metal phase transition occurs slightly above room temperature (at about 67°C) and can be induced thermally, electrically or optically [38, 39]. In [40] the authors perform electrical tuning of a plasmonic metasurface based on bow-tie antennas having the gap filled with VO_2 showing remarkable performances, however they do not provide general guidelines for the design of PCMs based resonators. To do so, in the following, we consider a plasmonic configuration: a

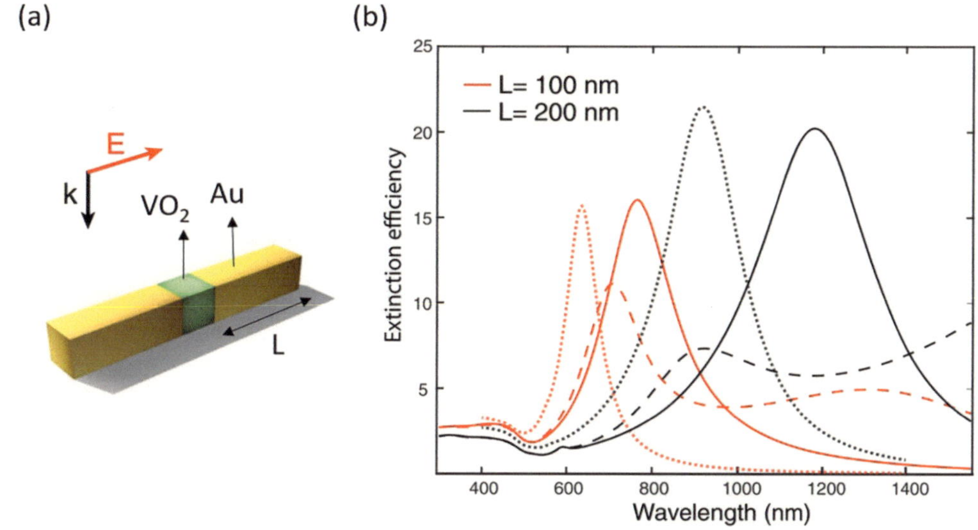

Figure 1.
(a) Sketch of the designed structure made by two gold arms with a gap filled with VO_2. (b) Extinction efficiency as a function of wavelength for different arm lengths. The continuous lines are associated with the PCM insulating phase; the dashed lines with the PCM metallic phase while the dotted lines correspond to the case where the gap is filled with air. The gap is fixed to 50 nm while the length L of the gold antenna is equal to 100 nm (red curves) and 200 nm (black curves), respectively.

single gold structure made of two arms with a square 50 × 50 nm^2 transverse section separated by a gap filled with VO_2, as shown in **Figure 1(a)** [41]. Let us highlight that the analysis with numerical simulations performed in Comsol Multiphysics v5.5— https://www.comsol.com— is useful to highlight some general considerations for the usage of PCMs in tunable optical systems which can be easily extended to other PCMs such as GeSbTe [42]. The proposed structure is placed in a homogenous surrounding material with refractive index equal to 1 (air). In the simulations, the gold nanoantenna is excited by a linearly polarized plane wave with electric field oriented along the long axis of the device, see **Figure 1**. We report – **Figure 1(b)** – the extinction efficiency for diverse arm length values above (dashed curves) and below (continuous curves) the critical transition temperature. For completeness, the case of air-filled gap is plotted in the figure with dotted curves. In the numerical investigation, we use the equilibrium optical constant at T = 30°C and T = 85°C for the insulating and metallic states, respectively. Few main aspects can be underlined by carefully observing the reported results.

Indeed, the optical properties are governed by the geometry of the structure. Firstly, when the arms lengths are increased, the resonances red-shift due to retardation effects, and broaden due to the increasing absorption of gold for wavelengths longer than 700 nm. Secondly, there is an evident arm length dependence of the shift between the resonances observed when the VO_2 inclusion is in either the metallic or the insulating phase. Therefore, the arm length represents a crucial parameter in these reconfigurable nanodevices. More specifically, the shift is smaller for shorter arm lengths. Moreover, the gap size can also affect the optical behavior. To investigate this dependence, in **Figure 2(a, b)** we report the extinction efficiency as a function of wavelength and gap size for the insulating and metallic phase, respectively. In these simulations the arms lengths are fixed to 200 nm. Regarding the insulating phase, we can notice a blue-shift of the resonant peak when the gap size is increased. This can be

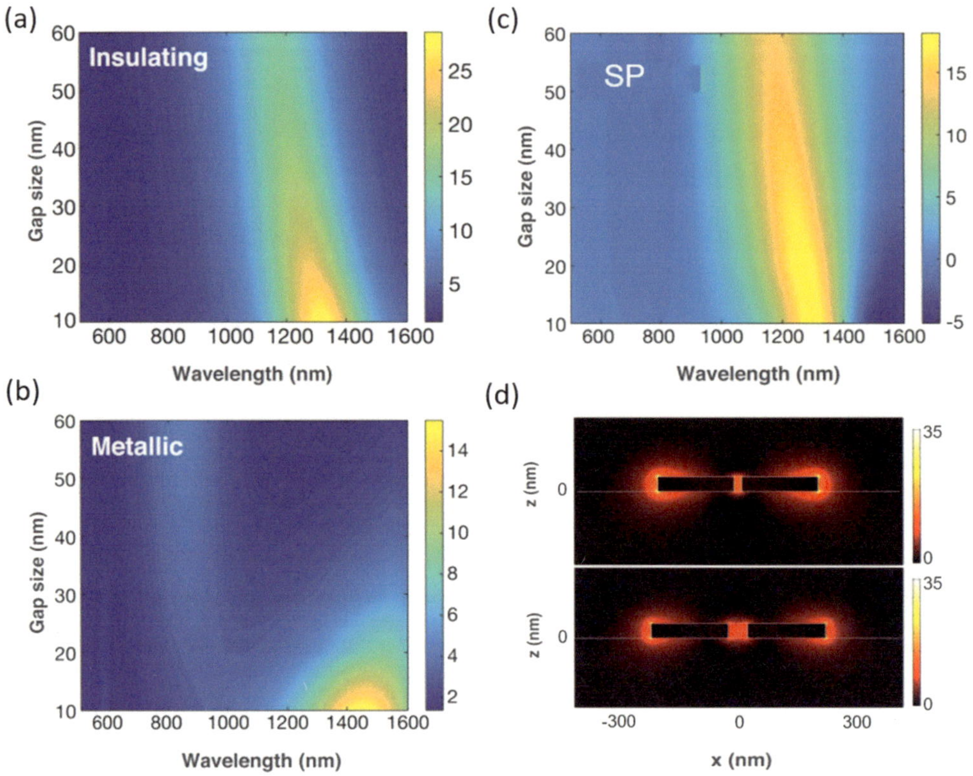

Figure 2.
Extinction efficiency σ_{ext} as a function of wavelength and gap size for the (a) insulating and (b) metallic phase of the VO_2 gap inclusion. The antenna arm length is 200 nm. (c) Switching parameter $\sigma_{ext}^{ins} - \sigma_{ext}^{met}$ as a function of wavelength and gap size. (d) Electric field enhancement ($|e|/E_0$) distribution at resonance for a structure with gap equal to 20 (top panel) and 50 nm (bottom panel).

understood by considering that, when the gap size is large enough, the two arms act as independent antennas. On the contrary, in the metallic state, a red shift of the less intense resonant peak as a function of the gap size can be seen. Intuitively, this can be explained by the fact that, when VO_2 is in the metallic phase, the entire device can be seen as a single antenna with increased length. To summarize all these results, we define the Switching Parameters, SP, as:

$$SP = \sigma_{ext}^{ins} - \sigma_{ext}^{met} \tag{1}$$

where σ_{ext}^{ins} and σ_{ext}^{met} represent the extinction efficiency for the VO_2 insulating or metallic phase, respectively. A plot of SP as a function of gap size and wavelength is reported in **Figure 2(c)**. The highest SP value is reached for small gaps, which better confine the electromagnetic field. **Figure 2(d)** elucidates the electric field distribution in the insulating phase for two different gap size values: 20 and 50 nm. One can clearly see that the electric field inside the gap is more uniform for the smaller value.

The SP can be evaluated also in the case of an asymmetric gap position. **Figure 3** shows the results obtained for this situation. The numerical predictions reveal that the maximum SP is achieved in the symmetric configuration, i.e., when the two arms are equal. Notably, the SP is not dramatically affected by small variations of

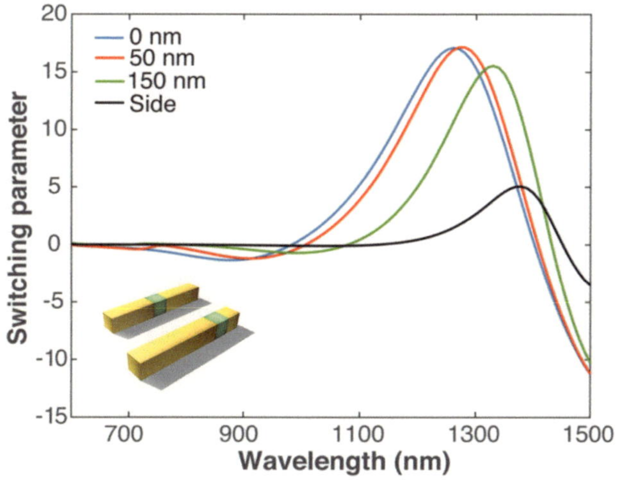

Figure 3.
Switching parameter as a function of the wavelength for different positions of the gap. The sketch represents the two conditions referred to in the legend as 0 nm (top) and 150 nm (bottom). In the legend, the number indicates the shift of the gap from the origin (center of the structure). The total length of the arms is 400 nm.

the gap position with respect to the perfectly symmetric condition. This means that the proposed device is robust with respect to fabrication defects that can alter the optimal geometrical design condition. Another important result is that the SP is not proportional to the volume of VO_2. This is of paramount importance since it allows to reduce the size of the device without reducing its functionality. Indeed, some recent works on PCM metasurfaces proved experimentally that a thin VO_2 film underneath dielectric resonators is sufficient to tune the behavior of a metasurface [43] and a metasurface of VO_2 nanocylinders can be employed to perform optical limiting [44].

To conclude, we have reported recent a recent example of a nanoresonator designed to maximize the metal–VO_2 near-field coupling to obtain tunable optical devices. The presented achievements may open new strategies for the implementation of optical structures whose behavior can be modulated by tuning external parameters, such as temperature, bias voltage and optical pump intensity [45, 46]. A metasurface composed by a repetition of the reported structure may benefit for even higher optical efficiency due to the resonance narrowing. Moreover, in the near future, this preliminary study related to the linear integration between nano-resonators and PCMs will also represent a guideline for nonlinear applications. For instance, in [47] the authors use the concept of PCM in the nonlinear regime and they numerically prove the tuning of second- harmonic generation (SHG) by gold metasurface made of split- ring resonators with a gap filled with GeSbTe alloy.

3. Liquid crystal-embedded nonlinear dielectric metasurface

In this section, we present recent results concerning the modulation of the nonlinear optical emission from dielectric metasurfaces excited in the infrared spectral region. As already mentioned before, despite the rapid advancements of nano-photonics and nanofabrication techniques, most of the up to now designed meta-structures exhibit a static behavior which is dictated by the fabrication process and cannot be changed [48–55]. However, the strong light–matter interaction in dielectric materials can pave the way for a dynamic control of the optical response via the modification of their electromagnetic properties [56]. So far, different tuning mechanisms have been proposed, ranging from electrical to mechanical and thermal control [29, 32, 57] and as well PCM [58, 59]. In this context, the first demonstrations were carried out by considering the metasurface as a passive object where the output light wavelength is the same as the incoming one [60, 61]. However, dielectric materials such as GaAs and AlGaAs have also a strong nonlinear response of the second order with higher efficiency with respect to plasmonic nano-antenna [62]. In such high-refractive index materials it is thus possible to efficiently generate a signal at a different frequency (i.e. the double) than the input laser excitation [63, 64]. In this way, the metasurface acts already as an active device.

Recently, liquid crystals (LCs) emerged as excellent candidates for achieving dynamic light manipulation at the nanoscale mainly thanks to their high birefringence that can be governed by temperature or by an external electric field. Linear dielectric metasurfaces embedded in a LC matrix have been successfully implemented [65–67] where the molding of the light scattered from the structure depends on the orientation of the LC axis. However, the reconfigurable control of the nonlinear harmonic light emitted by metasurfaces embedded in LCs is still preliminary [68].

In the following, we review one possible implementation to obtain a SHG modulator by using a commercial LC as the immersion material of a metasurface made of AlGaAs nanodiscs placed over a low-refractive index substrate.

Let us consider the metasurface depicted in **Figure 4(a)** [69]. The structure is a periodic array of nanodiscs of height equal to 200 nm. The AlGaAs crystal axes are oriented as sketched in the inset of **Figure 4(b)**. The dielectric structure is covered by a LC whose director is kept in the plane of the metasurface. The employed LC is E7 which is quite popular in the display industry [70–72].

To demonstrate the modulation capability of the proposed device, nonlinear numerical simulations have been carried out. More specifically, the nonlinear SH signal coming from the metasurface has been calculated as a function of the lattice period p, the radius r of the nanocylinder and the polarization of the incident pump beam. The calculations have been performed using Comsol Multiphysics. The

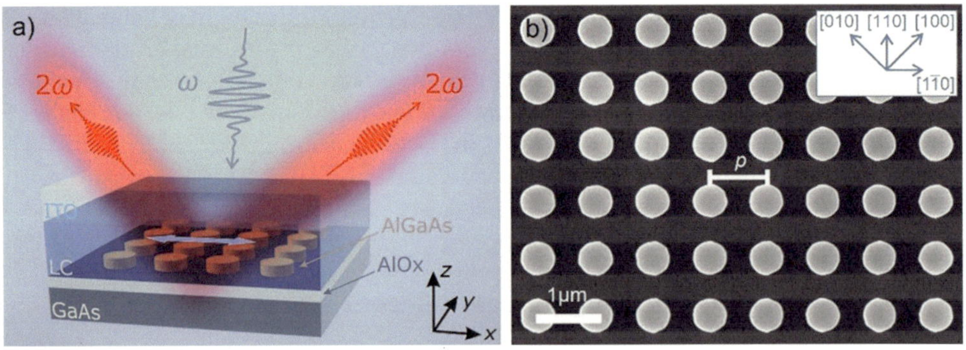

Figure 4.
(a) Sketch of the designed metasurface excited by the fundamental beam at ω. the unitary cell is constituted by a AlGaAs nanodisc that is placed on an Al_2O_3-GaAs substrate. An ITO superstrate is placed in view of future voltage-controlled implementations. The light blue arrow indicates the LC director parallel to the x axis. The harmonic generated at 2ω is measured as a function of the incident polarization. (b) Scanning electron microscopy image of the fabricated metasurface, with period p equal to 910 nm. Inset: The orientation of the AlGaAs crystalline axes.

AlGaAs nanodiscs are layered over an Al_2O_3 substrate. The refractive indices of the materials are the same as in Ref. [68]. For the modeling of the E7 LC we consider a homogeneous anisotropic dielectric material with ordinary (n_o) and extraordinary refractive index (n_e), respectively equal to 1.5 and 1.6 at the considered wavelength. Please note that this values slightly differ from the typically reported ones to account for imperfections in the device fabrication that lead to a lower anisotropy. In this way, the numerical results can mimic the realistic experimental conditions. The impinging beam is assumed to be a plane wave at normal incidence with a wavelength of 1551 nm. The SH emission is simulated in two steps. First, the field at the fundamental incident wavelength is evaluated. Then, the SH sources are computed in terms of current densities. More specifically, for the zincblende crystalline structure of AlGaAs the current density J_i can be computed as:

$$J_i(2\omega) = j\,\omega_{SH}\varepsilon_0\chi^{(2)}{}_{ijk}E_j(\omega)E_k(\omega) with i \neq j \neq k, \quad (2)$$

where i,j,k indicate the Cartesian axes, ε_0 the vacuum permittivity, $Ej(\omega)$ the jth component of the electric field at the fundamental frequency ω, $\chi^{(2)}{}_{ijk}$ the second-order susceptibility that we fix to 200 pm/V [64]. Please note that the nonlinear current densities are defined only in the dielectric medium because both the Al_2O_3 substrate and the LC have negligible $\chi^{(2)}$.

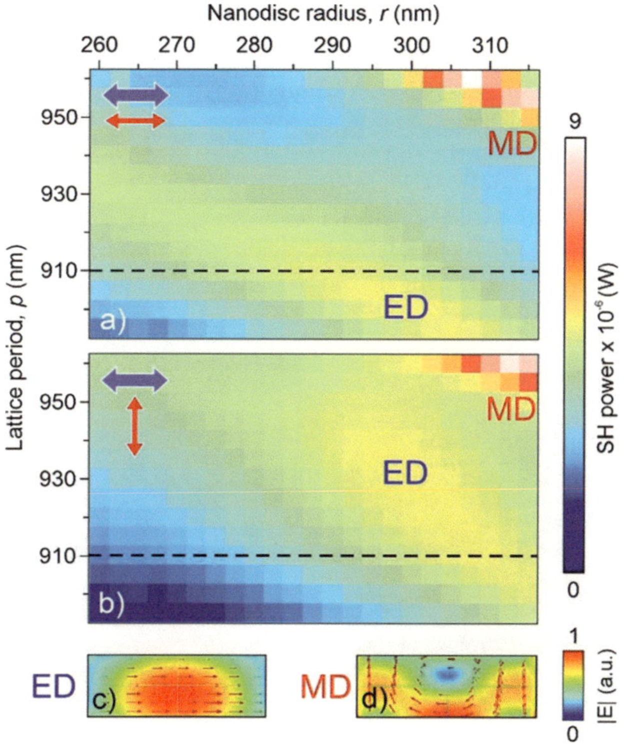

Figure 5.
Simulated SH emitted power as a function of metasurface period and nanodisc radius for linear pump polarization (red arrows) (a) parallel or (b) orthogonal to the LC director (blue arrows). Electric field distribution at (c) the electric (ED) and (d) magnetic dipolar (MD) resonance inside the nanodisc, x-z plane.

The obtained results are reported in **Figure 5**, where the predicted SH power maps computed in the backward direction are reported for the two considered incident beam polarizations, indicated as red arrows. In the simulations, the metasurface period p varies from 895 nm to 960 nm and the radius of the pillar r from 260 nm to 315 nm. **Figure 5(a)** and **(b)** highlight a strong dependence of the SH on the incident polarization. In particular, the numerical predictions reveal that the SH stem from an electric dipolar (ED) or magnetic dipolar resonance (MD) at the pump wavelength. The ED resonance is broader than the MD one, but it displays a more pronounced dependence on the relative orientation between the incident polarization and the LC director.

For instance, one can notice that, for p = 910 nm, (black dotted line in **Figure 5**), the ED resonances for the two incident light polarizations are observed at radii that differ by roughly 20 nm. Hence, the ED resonance is strongly modulated by the relative orientation between the incident light and the LC director. The latter is a crucial aspect for obtaining a metasurface that acts as a SH modulator. This promising consideration motivated the fabrication of such a device, as described in [69]. Briefly, the structure is fabricated by molecular-beam epitaxy on a non-intentionally doped GaAs wafer, with 200 nm layer of $Al_{0.18}Ga_{0.82}As$ on the top of an aluminum-rich substrate, which is later oxidized. Metasurfaces with varying cylinder radius in the range 260–315 nm and with period fixed at 910 nm have been obtained, as already reported in **Figure 4**. The metasurfaces are then embedded in a 12 μm-thick matrix of E7 LC following the procedure reported in [69].

The structures are then experimentally tested with the set-up already described in [64]. The source is an ultrafast laser with wavelength centered at 1551 nm which provides 160 fs pulses at 80 MHz repetition rate. The average excitation power is 10 mW that is associated with an average intensity on the sample of roughly 2 KW/cm^2 (peak intensity equal to 160 MW/cm^2) given a beam diameter of 25 μm. Please note that the metasurface acts as a diffraction grating for the SH signal. In particular, the first diffraction orders are supposed to be emitted with an angle of about 58° (accordingly to the Bragg's diffraction law).

The numerical aperture (NA) of the collection objective is equal to 0.85, hence only the first diffraction orders are detected by our set-up. The experimental Back-Focal Plane (BFP) images of the measured SH signal are reported in **Figure 6** as a

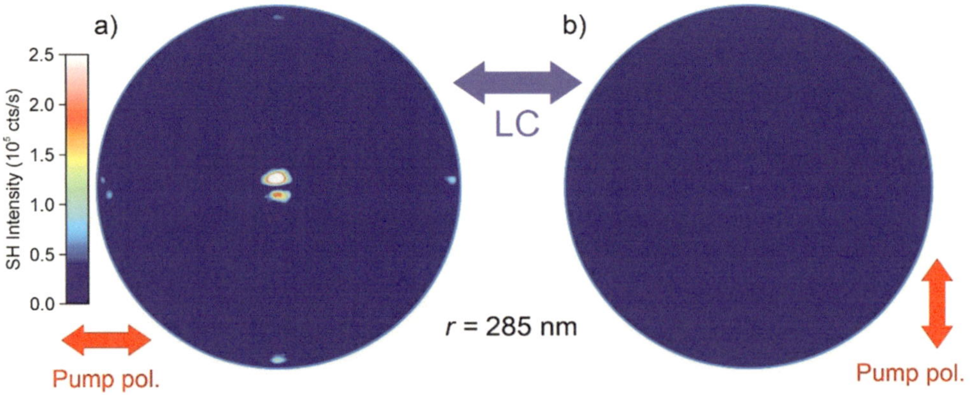

Figure 6.
Experimental BFP SHG maps of the proposed metasurfaces for the pump beam polarized (a) parallel or (b) orthogonal to the LC director. The red arrows represent the pump polarization while the blue arrow indicates the LC orientation.

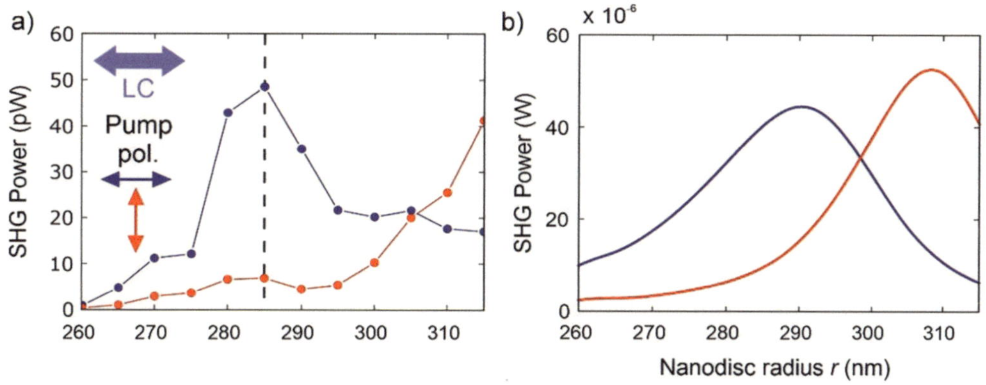

Figure 7.
Comparison between (a) experimental SH and (b) theoretical predictions as a function of the nanodisc radius for pump polarized parallel (blue line) or orthogonal (red line) to the LC director (light blue arrow). The vertical dashed black line marks the value of the nanodisc radius of the metasurface from which the BFP images reported in the previous figure were obtained.

function of the incident light polarization. Let us stress that the SH emission related to the first diffraction orders is at the edge of the NA and hence is partially cut off. However, it is clearly visible that the SH power is higher when the pump polarization is parallel to the LC director, see **Figure 6(a)**.

To validate and confirm this behavior, we acquired different BFP images by varying the nanodisc radius and the polarization of the incident light. **Figure 7** reports the obtained results compared to the numerical predictions. A shift in the SH peak is clearly visible when the incident polarization is switched from parallel (blue curve) to perpendicular (red curve) to the LC director. The experimental results and the theoretical predictions are in good agreement. We attributed the SH power trend to the shift of the ED resonance at the fundamental wavelength for the two considered excitations, see **Figure 5**. Importantly, for a metasurface with radius r = 285 nm the modulation of the SH power reaches the high value of about one order of magnitude, as highlighted by the dashed black line in **Figure 7**. In [69] the authors reported another structure where the LC director is always parallel to the (110) crystallographic axis. However, for the latter metasurface no significant SH modulation is expected and we will not further discuss this case here.

To summarize, for the first time to our knowledge, we have experimentally proven the modulation of the SH signal coming from an optimized dielectric metasurface covered by a liquid crystal matrix as a function of the incident pump beam polarization. Our experimental validation is achieved considering a fixed liquid crystal orientation (i.e. planar alignment). By switching the pump polarization, the refractive index felt by the meta-atoms as the upper medium modifies due to the anisotropic LC permittivity matrix. This variation shifts the wavelength position of the excited Mie resonances inside the dielectric and consequently the SH signal.

Importantly, for fixed incident wavelength and geometrical parameters of the metasurface, we obtain an order of magnitude enhancement in the emitted SH power when the incident pump polarization is switched from perpendicular to parallel to the LC director. Our experimental measurements, which are well reproduced by the numerical predictions, pave the way for the realization of nonlinear modulators in a scenario where the pump polarization is kept fixed and the liquid crystal anisotropy is controlled electrically or optically.

4. Conclusions

To conclude, in this chapter we have briefly discussed some of the recent progress in the field tunable nanophotonics. In particular, we focused on two recent approaches to attain efficient tuning of the meta-devices. We first discuss theoretical predictions about a specific phase-change material, VO_2, and how it can be used as a building block for reconfigurable meta-atoms. We have then reviewed the possibility of embedding dielectric metasurfaces in a liquid crystal matrix to achieve modulation of the nonlinear emitted light signal. These results clearly demonstrate how the control of the electromagnetic radiation can be achieved by well-optimized tunable metasurfaces. For practical applications, more efforts and solutions are needed for integrating reconfigurable metasurfaces within conventional electro-optical devices. The reduction of the fabrication costs is a key aspect to consider for a wide spreading of this technology. The versatility of nonlinear metasurfaces will certainly boost novel implementations in this research area.

Conflict of interest

The authors declare no conflict of interest.

Author details

Davide Rocco[1*], Andrea Locatelli[1], Domenico De Ceglia[1], Andrea Tognazzi[2], Attilio Zilli[3], Michele Celebrano[3], Marco Finazzi[3], Antonio Ferraro[4], Roberto Caputo[5] and Costantino De Angelis[1]

1 Department of Information Engineering, University of Brescia, Brescia, Italy

2 Department of Engineering, University of Palermo, Palermo, Italy

3 Department of Physics, Politecnico di Milano, Milano, Italy

4 Consiglio Nazionale delle Ricerche – Istituto di Nanotecnologia, Rende, Italy

5 Department of Physics, University of Calabria, Rende, Italy

*Address all correspondence to: davide.rocco@unibs.it

References

[1] Kogelnik H. Imaging of optical modes-resonators with internal lenses. Bell System Technical Journal. 1965;**44**:455-494

[2] Luo X. Subwavelength artificial structures: Opening a new era for engineering optics. Advanced Materials. 2019;**31**:1804680

[3] Courjon D, Bainier C. Near field microscopy and near field optics. Reports on progress in Physics. 1994;**57**:989

[4] Eldada L. Optical communication components. Review of Scientific Instruments. 2004;**75**:575-593

[5] Frazão O, Santos JL, Araujo FM, Ferreira LA. Optical sensing with photonic crystal fibers. Laser & Photonics Reviews. 2008;**2**:449-459

[6] Deng Z-L, Li G. Metasurface optical holography. Materials today. Physics. 2017;**3**:16-32

[7] Golubenko G, Svakhin A, Sychugov V, Tishchenko A. Total reflection of light from a corrugated surface of a dielectric waveguide. Soviet Journal of Quantum Electronics. 1985;**7**:886

[8] Bolotovskiĭ B, Stolyarov S. Reflection of light from a moving mirror and related problems. Soviet Physics Uspekhi. 1989;**32**:813

[9] Blodgett KB. Use of interference to extinguish reflection of light from glass. Physical Review. 1939;**55**:391

[10] Reed G, Mashanovich G, Gardes F, Thomson D. Silicon optical modulators. Nature Photonic. 2010;**4**:518-526

[11] Drezet A, Cyriaque G, Ebbesen T. Miniature plasmonic wave plates. Physical Review Letters. 2008;**101**: 043902

[12] Hatakoshi GI, Tanaka SI. Grating lenses for integrated optics. Optics Letters. 1978;**2**:142-144

[13] Bukhari S, Vardaxoglou JY, Whittow W. A Metasurfaces review: Definitions and applications. Applied Sciences. 2019;**9**:2727

[14] Ding F, Pors A, Bozhevolnyi S. Gradient metasurfaces: A review of fundamentals and applications. Reports on Progress in Physics. 2017;**81**:026401

[15] Imani M, Gollub J, Yurduseven O, Diebold A, Boyarsky M, Fromenteze T, et al. Review of Metasurface antennas for computational microwave imaging. IEEE Transactions on Antennas and Propagation. 2020;**68**:1860-1875

[16] Hsiao H, Chu C, Tsai D. Fundamentals and applications of Metasurfaces. Small Methods. 2017;**1**:1600064

[17] Li A, Singh S, Sievenpiper D. Metasurfaces and their applications. Nanophotonics. 2018;**7**:989-1011

[18] Alaee R, Albooyeh M, Rockstuhl C. Theory of metasurface based perfect absorbers. Journal of Physics D: Applied Physics. 2017;**50**:503002

[19] Sung J, Lee G, Lee B. Progresses in the practical metasurface for holography and lens. Nano. 2018;**8**:1701-1718

[20] Qiu C, Zhang T, Hu G, Kivshar Y. Quo vadis, metasurfaces? Nano Letters. 2021;**21**:5461-5474

[21] Jiang Q, Jin G, Cao L. When metasurface meets hologram: Principle

and advances. Advances in Optics and Photonics. 2019;**11**:518-576

[22] Chen S, Liu W, Li Z, Cheng H, Tian J. Metasurface-empowered optical multiplexing and multifunction. Advanced Materials. 2020;**32**:1805912

[23] Luo X. Subwavelength optical engineering with metasurface waves. Advanced Optical Materials. 2018;**6**:1701201

[24] Yue F, Wen D, Xin J, Gerardot B, Li J, Chen X. Vector vortex beam generation with a single plasmonic metasurface. ACS Photonics. 2016;**3**:1558-1563

[25] Ma X, Pu M, Li X, Huang C, Wang Y, Pan W, et al. A planar chiral meta-surface for optical vortex generation and focusing. Scientific Reports. 2015;**5**:1-7

[26] Camacho-Morales R, Rocco D, Xu L, Gili V, Dimitrov N, Stoyanov L, et al. Infrared upconversion imaging in nonlinear metasurfaces. Advanced Photonics. 2021;**3**:036002

[27] Roy T, Zhang S, Jung I, Troccoli M, Capasso F, Lopez D. Dynamic metasurface lens based on MEMS technology. APL Photonics. 2018;**3**(021302):2018

[28] Shlezinger N, Alexandropoulos G, Imani M, Eldar Y, Smith D. Dynamic metasurface antennas for 6G extreme massive MIMO communications. IEEE Wireless Communications. 2021;**28**: 106-113

[29] Celebrano M, Rocco D, Gandolfi M, Zilli A, Rusconi F, Tognazzi A, et al. Optical tuning of dielectric nanoantennas for thermo-optically reconfigurable nonlinear metasurfaces. Optics Letters. 2021;**46**:2453-2456

[30] Rocco D, Gandolfi M, Tognazzi A, Pashina O, Zograf G, Frizyuk K, et al. Opto-thermally controlled beam steering in nonlinear all-dielectric metastructures. Optics Express. 2021;**29**:37128-37139

[31] Carletti L, de Ceglia D, Vincenti MA, De Angelis C. Self-tuning of second-harmonic generation in GaAs nanowires enabled by nonlinear absorption. Optics Express. 2019;**27**:32480

[32] Decker M, Kremers C, Minovich A, Staude I, Miroshnichenko A, Chigrin D, et al. Electro-optical switching by liquid-crystal controlled metasurfaces. Optics Express. 2013;**21**:8879-8885

[33] Schirato A, Mazzanti A, Zaccaria R, Nordlander P, Alabastri A, Della VG. All-optically reconfigurable plasmonic metagrating for ultrafast diffraction management. Nano Letters. 2021;**21**:1345-1351

[34] Pogna E, Celebrano M, Mazzanti A, Ghirardini L, Carletti L, Marino G, et al. Ultrafast, all optically reconfigurable, nonlinear nanoantenna. ACS Nano. 2021;**15**:11150-11157

[35] Yuan L, Fan S. Temporal modulation brings metamaterials into new era. Light: Science & Applications. 2022;**11**:1-2

[36] Appavoo K, Haglund RF. Polarization selective phase-change nanomodulator. Scientific Reports. 2014;**4**:1-6

[37] Wang H, Yang Y, H. and Wang L. Switchable wavelength-selective and diffuse metamaterial absorber/emitter with a phase transition spacer layer. Applied Physics Letters. 2014;**105**:071907

[38] Qazilbash Q, Brehm M, Chae BG, Ho PC, Andreev GO, Kim BJ, et al. Mott transition in VO2 revealed by infrared spectroscopy and nano-imaging. Science. 2007;**318**:1750-1753

[39] Kana JBK, Ndjaka JM, Vignaud G, Gibaud A, Maaza M. Thermally tunable optical constants of vanadium dioxide thin films measured by spectroscopic ellipsometry. Optics Communications. 2011;**284**:807-812

[40] Zhu Z, Evans P, Haglund R, Valentine J. Dynamically reconfigurable metadevice employing nanostructured phase-change materials. Nano Letters. 2017;**17**:4881-4885

[41] Tognazzi A, Locatelli A, Vincenti M, Giannetti C, De Angelis C. Tunable optical antennas using vanadium dioxide metal-insulator phase transitions. Plasmonics. 2019;**14**:1283-1288

[42] de Galarreta C, Sinev I, Alexeev A, Trofimov P, Ladutenko K, Carrillo SG, et al. Reconfigurable multilevel control of hybrid all-dielectric phase-change metasurfaces. Optica. 2020;**7**:476-484

[43] Tripathi A, John J, Kruk S, Zhang Z, Nguyen H, Berguiga L, et al. Tunable Mie-resonant dielectric Metasurfaces based on VO2 phase-transition materials. ACS Photonics. 2021;**8**:1206-1213

[44] Howes A, Zhu Z, Curie D, Avila J, Wheeler V, Haglund R, et al. Optical limiting based on Huygens' metasurfaces. Nano Letters. 2020;**20**:4638-4644

[45] Seo G, Kim B, Choi J, Lee Y, Kim H. Direct current voltage bias effect on laser-induced switching bistability in VO2-based device. Applied Physics Express. 2012;**5**:102201

[46] Tao Z, Zhou F, Han T, Torres D, Wang T, Sepulveda N, et al. The nature of photoinduced phase transition and metastable states in vanadium dioxide. Scientific Reports. 2016;**6**:1-10

[47] Guo K, Zhou K, Guo Z. Tunable second harmonic generation from bianisotropic plasmonic metamolecule via utilizing phase change materials. Journal of Applied Physics. 2020;**128**:133104

[48] Overvig AC, Shrestha S, Malek SC, Lu M, Stein A, Zheng C, et al. Dielectric metasurfaces for complete and independent control of the optical amplitude and phase. Light: Science and Applications. 2019;**8**:1-12

[49] Liu W, Li Z, Cheng H, Chen S. Dielectric resonance-based optical metasurfaces: From fundamentals to applications. Iscience. 2020;**23**:101868

[50] Genevet P, Capasso F, Aieta F, Khorasaninejad M, Devlin R. Recent advances in planar optics: From plasmonic to dielectric metasurfaces. Optica. 2017;**4**:139-152

[51] Albella P, Poyli M, Schmidt M, Maier S, Moreno F, Saenz J, et al. Low-loss electric and magnetic field-enhanced spectroscopy with subwavelength silicon dimers. Journal of Physical Chemistry C. 2013;**117**:13573-13584

[52] Shresta S, Overvig A, Lu M, Stein A, Yu N. Broadband achromatic dielectric metalenses. Light: Science Applications. 2018;**7**:1-11

[53] Zhao R, Xiao X, Geng G, Li X, Li J, Li X, et al. Polarization and holography recording in real- and k-space based on dielectric metasurface. Advanced Functional Materials. 2021;**31**:2100406

[54] Guo Z, Zhu L, Shen F, Zhou H, Gao R. Dielectric metasurface based high-efficiency polarization splitters. RSC Advances. 2017;**7**:9872-9879

[55] Zhang Q, Li M, Liao T, Cui X. Design of beam deflector, splitters, wave plates and metalens using photonic elements with dielectric metasurface. Optics Communication. 2018;**411**:93-100

[56] He Q, Sun S, Zhou L. Tunable/reconfigurable metasurfaces: Physics and applications. Research. 2019;**2019**:1-16

[57] Gutruf P, Zou C, Withayachumnankul W, Bhaskaran M, Sriram S, Fumeaux C. Mechanically tunable dielectric resonator metasurfaces at visible frequencies. ACS Nano. 2016;**10**:133-141

[58] Wang Y, Landreman P, Schoen D, Okabe K, Marshall A, Celano U, et al. Electrical tuning of phase-change antennas and metasurfaces. Nature Nanotechnology. 2021;**16**:667-672

[59] Zhang Y, Fowler C, Liang J, Azhar B, Shalaginov M, Deckoff-Jones S, et al. Electrically reconfigurable non-volatile metasurface using low-loss optical phase-change material. Nature Nanotechnology. 2021;**16**:661-666

[60] Komar A, Fang Z, Bohn J, Sautter J, Decker M, Miroshnichenko A, et al. Electrically tunable all-dielectric optical metasurfaces based on liquid crystals. Applied Physics Letters. 2017;**110**:071109

[61] Komar A, Paniagua-Domınguez R, Miroshnichenko A, Yu Y, Kivshar Y, Kuznetsov A, et al. Dynamic beam switching by liquid crystal tunable dielectric metasurfaces. ACS Photonics. 2018;**5**:1742-1748

[62] Gigli C, Leo G. All dielectric X(2) metasurfaces: Recent progress. Opto-Electronic. Advances. 2022;**210093**:1

[63] Marino G, Rocco D, Gigli C, Beaudoin G, Pantzas K, Suffit S, et al. Harmonic generation with multi-layer dielectric metasurfaces. Nano. 2021;**10**:1837-1843

[64] Gili V, Carletti L, Locatelli A, Rocco D, Finazzi M, Ghirardini L, et al. Monolithic AlGaAs second-harmonic nanoantennas. Optics Express. 2016;**24**:15965-15971

[65] Bohn J, Bucher T, Chong K, Komar A, Choi D-Y, Neshev D, et al. Active tuning of spontaneous emission by Mie-resonant dielectric metasurfaces. Nano Letters. 2018;**18**:3461-3465

[66] Zou C, Komar A, Fasold S, Bohn J, Muravsky A, Murauski A, et al. Electrically tunable transparent displays for visible light based on dielectric metasurfaces. ACS Photonics. 2019;**6**:1533-1540

[67] Hu Y, Ou X, Zeng T, Lai J, Zhang J, Li X, et al. Electrically tunable multifunctional polarization-dependent Metasurfaces integrated with liquid crystals in the visible region. Nano Letters. 2021;**21**:4554-456221

[68] Rocco D, Carletti L, Caputo R, Finazzi M, Celebrano M, De Angelis C. Switching the second harmonic generation by a dielectric metasurface via tunable liquid crystal. Optics Express. 2020;**28**:12037-12046

[69] Rocco D, Zilli A, Ferraro A, Borne A, Vinel V, Leo G, et al. Tunable second harmonic generation by an all-dielectric diffractive metasurface embedded in liquid crystals. New Journal of Physics. 2022;**24**:045002

[70] Schadt M. Liquid crystal materials and liquid crystal displays. Annual Review of Materials Science. 1997;**27**:305-379

[71] Bahadur B. Liquid crystal displays. Molecular crystals and liquid crystals. 1984;**109**:3-93

[72] Li J, Wen C, Gauza S, Lu R, Wu S. Refractive indices of liquid crystals for display applications. Journal of Display Technology. 2005;**1**:51

Chapter 4

Metamaterial Applications in Modern Antennas

Mohamed Lashab, Mounir Belattar, Sekkache Hocine and Halfaya Ahmed

Abstract

The chapter presents different types of metamaterials, their historical evolution, physical properties, and applications in antennas design. Metamaterials are artificial structures that offer electric and magnetic properties that are not found in nature, such are negative permittivity and negative permeability. These properties are used in antennas design in order to obtain ultrawide bandwidth, high gain, and electrically small structures. Modern wireless mobile communication uses 5G technology and multi-input multi-output (MIMO) antennas, which are based on high-frequency transmission and ultrawide band. Metamaterials are very good candidate for this technology, where miniaturized antennas loaded with metamaterials structures are used. Recently, electromagnetic sensors were used for liquid identification in biological and medical substances based on metamaterials for the sake of high sensitivity and better classification. Different metamaterial types are used in these sensors, depending on the nature of liquid and the associated application.

Keywords: antennas, metamaterials, wideband, mobile, metamaterial applications

1. Introduction

Modern antenna design and fabrication are based on the suitable choice of the substrate material, technology of fabrication, and domain or type of application [1]. Generally, the choice of substrate material has an effect on the frequency range, losses reduction, and polarization mode or harmonics generation [2]. This chapter is concerned with the second and third point of antenna design, technology of fabrication, and the type or domain of application.

There are various techniques or technologies of fabrication of electromagnetic structure in general, and each one of them has specific interest and can be used for given application [3]. For instance, dielectric antennas resonators, known as DRA, are generally used to generate specific harmonics and propagation mode, on the top surface of the antenna, where various types of dielectrics are used [4]. The substrate-integrated waveguide (SIW) is a type of antenna where a waveguide is built on the top of the substrate, and the main advantages of these antenna are lower losses and better power handling [4, 5]. The metamaterial antennas (MTMs) are based on artificial materials that have uncommon physical properties, such as negative permittivity or negative permeability; these types of antennas have many advantages, such as higher gain, wideband of frequency, and miniaturization effect [6].

Antenna design depends also on the domain of application, in this chapter electromagnetic sensors loaded with metamaterial are treated, and the higher sensitivity is obtained for dielectric identification of the ethanol liquid [7]. Compensated right-left hand (CRLH) structure is a special type of MTM, which is used to improve the sensitivity, and other structures of metamaterial are also used [8].

Metamaterials are artificial materials that are not found in nature, they are fabricated by special design of copper on any dielectric substrate [4]. Based on the principle of transmission lines (TLs), the early researchers have worked hard to find the first type of MTM known as split ring resonator (SRR), CSRR as complementary split ring resonator, and CLL as capacitive-loaded loop [9].

2. Historical development of metamaterials

The history of metamaterials started from artificial dielectric fabrication as microwave structures, and they were manufactured just after the Second World War [10], although the exploration of the artificial materials has started by the end of the nineteenth century. The first studies on metamaterials started in 1904, and these studies were carried out until the twentieth mid-century; all the research work done on this period was dealing with the phase velocity, group velocity, the pointing vector, and the wave vector [11]. In 1952, Kock, Rotman, and Schelkunoff started working on artificial dielectric [12], and they suggested magnetic particles made by capacitive-charged loops, the research aimed to obtain very high permeability. Further research work led to obtain completely modified dielectric, with desired permittivity and desired permeability; hence, the medium is considered as metamaterial [13].

Horace Lamb and Arthur Schuster [14] in their theoretical study approved the possibility of negative phase velocity accompanied with anti-parallel group velocity, and unfortunately, the experimental work was not done. In 1945, Leonid Mandelstam studied in detail the anti-parallel phase and the velocity group and proved possibility of negative velocity group in crystalline medium.

The first theoretical model of so-called metamaterial medium was examined by Veselago in 1967; due to lack of suitable materials and technical problems, the experimental work was delayed till the end of 1990. Veselago was the first to predict the reversed electromagnetic phenomena including the index of refraction; he invented the term "left-hand material" and studied the material with double-negative dielectric parameters, permeability, and permittivity; he published his famous paper in 1968 and gained the Nobel Prize [15].

John Pendry in 2000 presented the first practical research work based on metallic periodic structures as concentric shapes, called split ring resonator (SRR); his experimental results confirmed the theory proposed by Veselago. Indeed, he obtained negative permeability and negative permittivity around the resonant frequency of the SRR, and the negative index of refraction was also obtained by making specific combination of copper wires and cells [16]. He succeeded in modifying the magnetic structure of any material, only by providing SRR structures on the top of the material.

In 2003, Caloz and Itoh presented their first work on metamaterials based on transmission lines (TLs) [17–19], and they introduced a new microwave technology, namely the zero-order resonators (ZORs), based on composite right-left hand (CRLH) structures, considered as metamaterials. The CRLH technology is based on a combination of right-hand (RH) and left-hand (LH) electric equivalent circuits, leading to a resonant frequency of the microwave structure independently of its dimension [19]. In 2006, Nader Engheta and Richard W. Ziolkowski [20] published their first investigation on physical aspect of metamaterials, and they presented theoretical development on different types of metamaterials, single-negative (SGN) and double-negative (DNG) medium, plus some applications on antennas.

3. Physical properties of metamaterials

3.1 Metamaterial definition

Metamaterials ("MTMs") are generally defined as homogeneous electromagnetic structures with unusual properties that are not available in nature [21]; homogeneous structure is also defined as structure with average size of the unit cell much smaller than the wavelength applied on this unit cell [22]. By taking into account this condition, the relationship that relates the effective parameters, known as the effective electric permittivity $\boldsymbol{\varepsilon_{eff}}$ and the effective magnetic permeability $\boldsymbol{\mu_{eff}}$ and the index of refraction n, is given by the following equation.

$$n = \sqrt{\pm \varepsilon_{eff} \mu_{eff}} \tag{1}$$

In all types of metamaterials, the value of the constitutive parameters permittivity and permeability depend on the mutual space between cells, also on the geometrical shape of the unit cell. The desired constitutive parameters are obtained once the cell shape and the spacing are predefined; generally, the cells are implemented on planar substrate, and the dielectric of the substrate can affect greatly the constitutive parameters [23].

Materials are generally classified into four categories as shown in **Figure 1**: the first region is for double-positive (DPS), where the permeability and the permittivity are

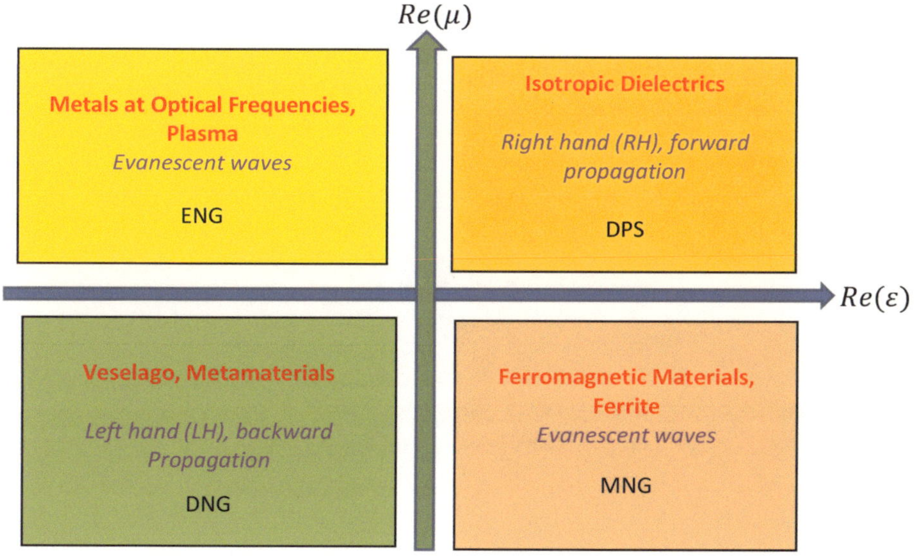

Figure 1.
Material classification.

both positive; this region is known for isotropic dielectrics, right-hand (RH) materials; the second region is for permittivity negative, ENG region, considered as metamaterial region, realized with parallel plates or wires; the third region is for double-negative parameters (DNG); both the permittivity and permeability values are negatives; this region is for materials known as left-hand (LH) materials.

The last region is for permeability negative (MNG), and this region is for ferro-magnetic and ferrite materials, generally fabricated from SRRs. In all regions near the zero axes, ENZ (ε near zero) and MNZ (μ near zero), both cases lead to ZIM (zero index metamaterial) [24].

3.2 Theoretical aspect of the SRR

Split ring resonators (SRRs) are artificial materials, and their permittivity and the permeability are not known; they vary with respect to the operating frequency, and the constitutive parameters are extracted either by experimental tests or by analytical models [25]. One of the well-known analytical methods is Drude-Lorentz model [26–28], known as the dispersion model, and this is a very accurate analytical method used to extract the constitutive parameters for unknown structures. For the split ring resonator or complementary SRR, **Figure 2**, using Drude-Lorentz model, the magnetic permeability and electric permittivity are described as follows:

$$\mu_r = \mu_o \left(1 + \frac{F_u f^2}{f^2 - f_{ou}^2 + j\gamma f}\right) \tag{2}$$

And,

$$\varepsilon_r = \varepsilon_o \left(1 + \frac{F_e f^2}{f^2 - f_{oe}^2 + j\gamma f}\right) \tag{3}$$

where ε_0 and μ_0 are respectively the background permittivity and the background permeability, whereas F_e and F_u are respectively the electric resonant intensity and the magnetic resonant intensity.

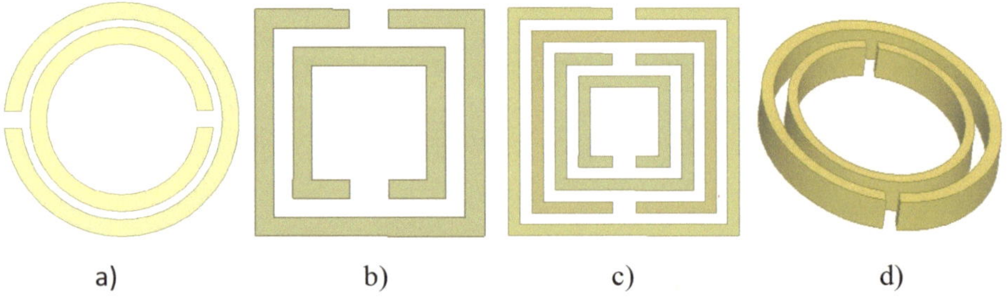

Figure 2.
Schematic representation of different types of elementary unit cell structures of SRRs: (a) double-circular SRR, (b) double-square SRR, (c) multi-square SRR, (d) double-cylindrical SRR.

We define also f_{oe} as the electric resonant frequency and f_{ou} the magnetic resonant frequency, the parameter γ is the damping factor of the resonance; several technics or algorithms are used to calculate all these parameters, for given frequencies [29].

3.3 CRLH resonator

The CRLH resonator is based on transmission line (TL) structures, which is essential to realize any resonant planar antenna with no dependence on its physical dimension, and this structure supports an infinite wavelength at its fundamental mode, which is required to realize ZOR resonators [30]. A practical realization of a left-hand (LH) material includes unavoidable right-hand (RH) effects, and the CRLH resonator is able to support an infinite wavelength ($\beta = 0$, with $\omega \neq 0$) and therefore, can be used to realize any planar microwave structures [31].

The equivalent circuit model of the CRLH unit cell is shown in **Figure 3**, as "T" shape. The resonance condition of such open-ended resonator is given by the dispersion relation given below [32].

$$\beta = \frac{n\pi}{l}(0, \pm 1, .., \pm (N-1)) \tag{4}$$

where "*l*" is the physical length of the resonator, *n* is the mode number, and *N* is the number of unit cells used to realize the CRLH, the equivalent electric circuit of a single unit cell is given by **Figure 3**, the electrical circuit contains an impedance Z and an admittance Y.

When the unit cell of zero-order resonator is inserted into the microwave structure, the size of this latter can be reduced since the resonant frequency does not depend on the physical dimension. In the case of open boundary conditions, the infinite wavelength resonance depends on the shunt parameters of the unit cell, the resonant frequencies are given by:

$$\omega_{sh} = \frac{1}{\sqrt{L_L C_R}} \text{ and } \omega_{ch} = \frac{1}{\sqrt{L_R C_L}} \tag{5}$$

where ω_{sh} is the shunt resonant frequency and ω_{ch} the serial resonant frequency [32]. More complicated zero-order structures than those given in **Figure 3** can be used, such as "Pi" shape or "T" shape, where the input and output have symmetrically the same components as in [33]; combination of the two types of shapes "T" and "Pi" can be employed to obtain more than two resonant frequencies [34].

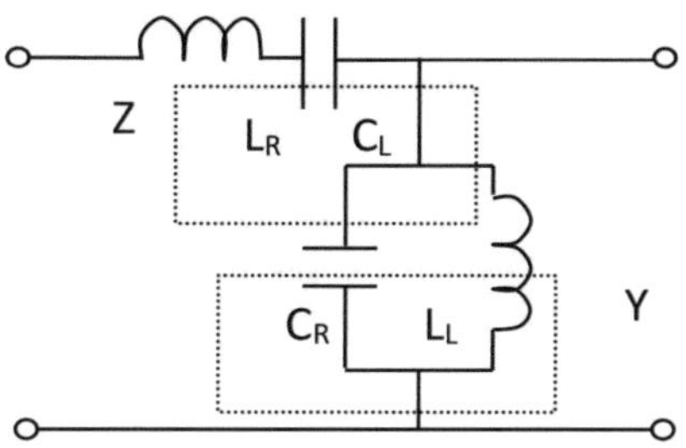

Figure 3.
Equivalent electric circuit of the proposed CRLH.

4. Antennas based on metamaterials

4.1 Antenna description

A monopole antenna loaded with CRLH resonator used for LTE and WiMax application is presented in this section, and the CRLH unit cell is designed by a zigzag resonator that is inserted between two symmetrical parts, with bow-tie shapes of the monopole antenna [33]. An inter-digital capacitor in terms of zigzag slot is also incorporated inside the lower part of the monopole antenna to support the miniaturization process, **Figure 4**. The proposed antenna exhibits the UWB and the miniaturization effect along with an improvement of the gain and the bandwidth subject to optimize the process of CRLH structure. The achieved results of the loaded monopole with CRLH show that the antenna is covering an operational bandwidth of 1.8–2.15 GHz as LTE band, 4.0–6.5 GHz as WiMax band, and 6.8–12.15 GHz as X band.

For more than 20 years ago, published research work on monopole antennas was mainly dealing with applications related to microwave devices, and this is mostly due to the low cost of fabrication of these types of antennas, and in addition, they were very basic. However, the early research studies on these antennas were not subject to improvement regarding their performances such as the high gain and large bandwidth [33]. More interesting research works published recently on monopole antennas loaded with artificial materials or dielectrics have shown that these antennas brought great improvement on the gain and bandwidth, which make them good candidates for modern communication devices [34]. Generally, antennas loaded with metamaterials have very important performances, but in the literature, it has been shown that antennas loaded with resonant structures such as the SRRs are lossy with narrow bandwidth and not suitable for microwave applications. For instance, LHM structures (left-hand metamaterials) based on TL (transmission line) have wider bandwidth and lower losses [18].

Microwave structures loaded with CRLH are generally based on the insertion of inter-digital capacitors and shorted stub inductors inside the antenna or the sensor, by suitable choice of the components dimensions, and one can have desired resonant frequency [30]. It has been shown that by increasing the shunt and inductors, the

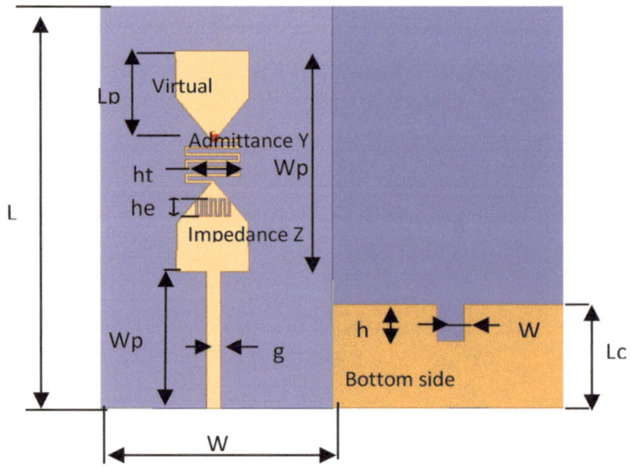

Figure 4.
Geometrical dimension of the proposed antenna, front and bottom.

antenna size can be made much smaller. In this proposed structure, the shunt inductance is accomplished by embedding a chip inductor on one of the strips of the monopole antenna, and thus, it omits the need for vias to the ground. Inter-digital capacitor and virtual ground have been widely used in radio frequency application; for instance, small resonators are one of these applications [31].

In the literature, most of the research works have shown that structures loaded with ZOR-TL are used to make any antennas or microwave sensors as electrically small, they will appear electrically large. ZOR-TL technique is used to improve the antennas matching and achieve good radiation properties or better microwave structures. This technique could be implemented on any antennas by producing a zero-phase constant for a nonzero frequency, which means the wavelength of the traveling wave becomes infinite on the ZOR-TL. This is the most important technique, which makes the resonance condition completely independent from the physical dimensions of the microwave structure [30, 32]. This technique can be used to design miniaturized antennas for any microwave structure, and the resonance of such microwave devices for any operating frequency depends only on the CRLH characteristics to acquire ZOR at that frequency and nothing to do with the physical dimensions of the antenna.

4.2 Results and discussion

The proposed antenna is presented in **Figure 4**, the substrate is an FR4 with relative permittivity 4.4 and dielectric loss tangent 0.02, the thickness of the substrate is 1.6 mm, and geometrical dimensions are given in **Table 1**. The proposed antenna is constituted by top metallic patches as the radiating parts plus a defected ground on the bottom side of the antenna, to achieve a good impedance matching to 50 Ω, and the proximity coupling is used as the feed network. Two parts are added as CRLH components: first, the zigzag strip considered as admittance Y, and an inter-digital capacitor considered as the Z impedance. The upper part of the antenna contains a virtual ground, whereas the down part contains a defected ground; by this description, the equivalent electric circuit of the monopole antenna is exactly the electric circuit given in **Figure 3**.

Several types of antenna structures were investigated as shown in **Figure 5**, and the figure shows three types of antennas: the original antenna, antenna 2, and antenna 3. The reflection coefficients of the proposed antenna including the three prototype antennas given in **Figure 5** are presented in **Figure 6**. The figure presents two resonant frequencies corresponding exactly to ω_sh and ω_ch in which these frequencies are independent from the antenna size. For all the prototype antennas, the first resonant frequency is around 5.5 GHz, and the second one is around 11.12 GHz; it is remarkable that these frequencies can be shifted just by changing the ZOR shape.

L	W	Lp	Wp	ht
40	26	9.5	15	6
Lc	he	Wt	h	g
11.55	2	3	4	1.5

Table 1.
Parameters of the proposed antenna in mm.

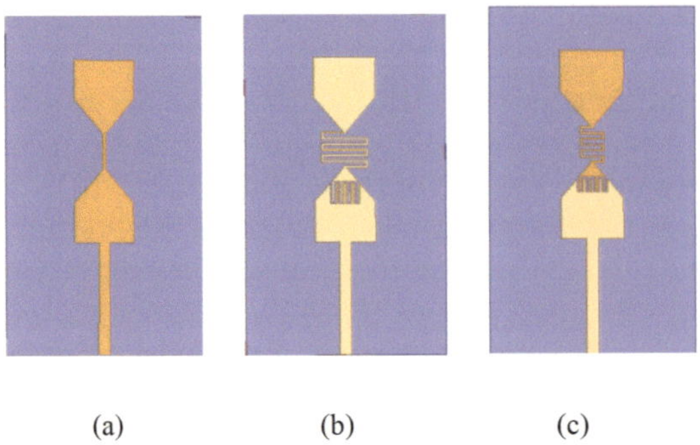

Figure 5.
Three types of antennas: (a) original antenna , (b) antenna 2, he = 3 mm, (c) Antenna 3, ht = 3 mm.

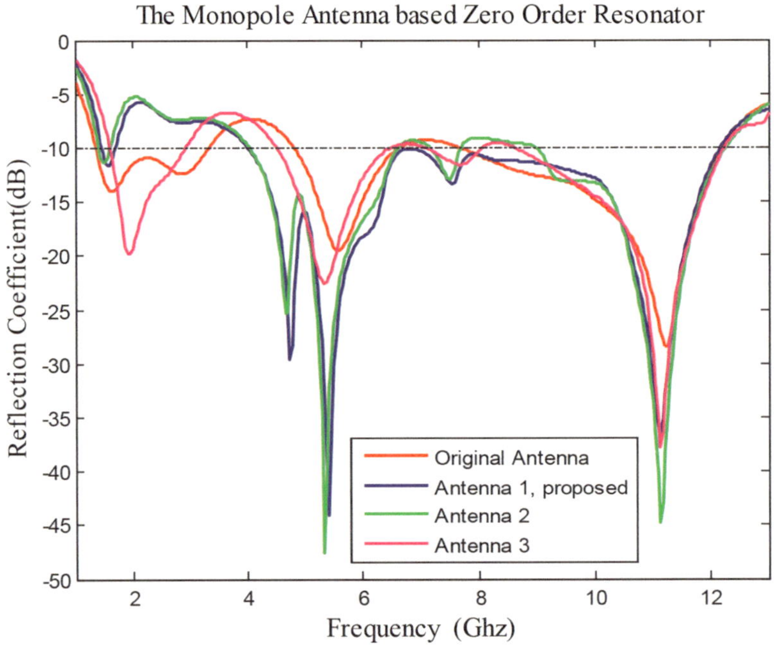

Figure 6.
Reflection coefficients for different types of antennas.

The bandwidth of the proposed antenna is 12.15GHz – 4GHz = 8.15 GHz, which is nearly 148% impedance bandwidth. Since the gain and radiation pattern of antenna 1 and antenna 2 are similar, we confirm that the variation of the capacitance value in antenna 2 is more significant than the variation of inductor length in antenna 3. The radiation pattern in the xz-plane given in **Figure 7** shows that there is an improvement for the case of antenna1 and antenna 2, which is from −3 dBi to 1.4 dBi at frequency 5.5 GHz compared with the original antenna. It was also noticed that there is a slight increase of gain values concerning antenna 3 from −3 dBi to −2.5 dBi.

The radiation pattern in the yz-plane given in **Figure 8** presents a very remarkable improvement of the gain values of antenna 1 and antenna 2 at 5.5 GHz, which is 1.8 dBi compared with the original antenna, that is, −3 dBi. Antenna 2 and antenna 1 have almost the same improvement in the gain, but antenna 1 has a larger bandwidth.

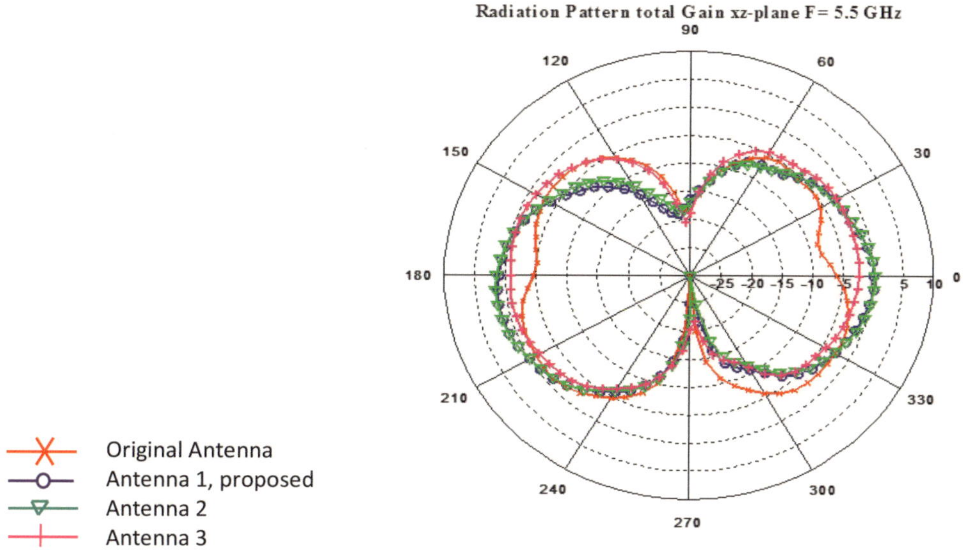

Figure 7.
Total gain radiation pattern in the xz-plane, F = 5.5 GHz.

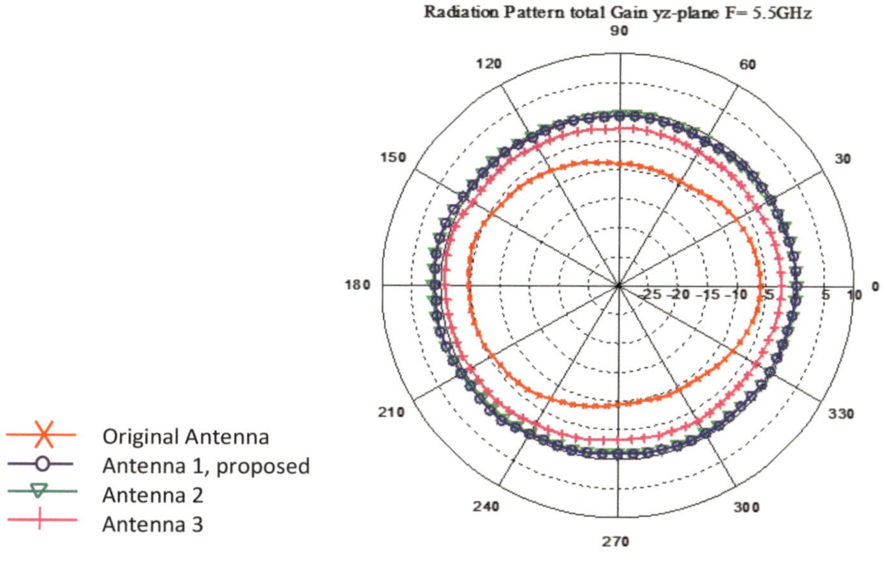

Figure 8.
Total gain radiation pattern in the yz-plane, F = 5.5 GHz.

5. Electromagnetic sensors based on metamaterial

5.1 Theoretical aspect

The main parts of the proposed sensor are the CRLH unit cell on the front side and the CSRR on the ground plane side, **Figure 9**. The sensor can be simulated and then optimized using the full-wave simulator Ansys HFSS.

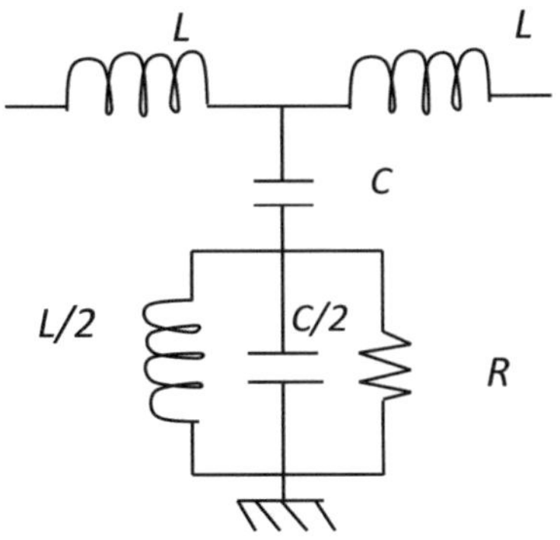

Figure 9.
Equivalent circuit of the proposed sensor, bottom view.

The equivalent electric circuit of the proposed sensor is the top side in **Figure 3**, whereas the equivalent electric circuit of the bottom side is given in **Figure 9**. In the CRLH unit cell, the impedance Z provides a series resonator based on (L_R, C_L), whereas Y admittance provides a parallel resonator based on (L_L, C_R) [30–38].

5.2 Geometrical dimensions of the sensor

The proposed sensor is presented in **Figure 10a**, the substrate is an FR4 with relative permittivity 4.4 and dielectric loss tangent 0.02, the thickness of the substrate is 1.6 mm.

The sensor is composed of CRLH and CSRR circuits with two ports, the sensor presents a coplanar structure, grounded at the bottom side, the bottom ground is first filled with PEC ground and then with CSRR, and this is given in **Figure 10b**.

The proximity coupling is used as the feed network to achieve a good impedance matching to 50 Ω, the geometrical dimensions are given in **Table 2**.

5.3 Results and discussion

The transmission coefficient of the sensor without liquid mixture (air) is presented in **Figure 11**, and the figure shows three types of liquid plus the air (without liquid).

The substances are ethanol, methanol, distilled water, and air, and their resonant frequencies are, respectively, 1.06 GHz, 1.01 GHz, 0.92 GHz, and 1.22 GHz. The aim of this work is to load the sensor with different volume fraction of ethanol, the measured resonant frequencies enable the calculation of the sensor sensitivity, and the transmission coefficient of different volume fraction is presented in **Figure 12**; in this figure the response of air is missing.

The frequency range of the sensor is from 1.065 GHz to 1.195 GHz, a difference is of 0.13 GHz. The transmission coefficient of the sensor loaded with CSRR is given in

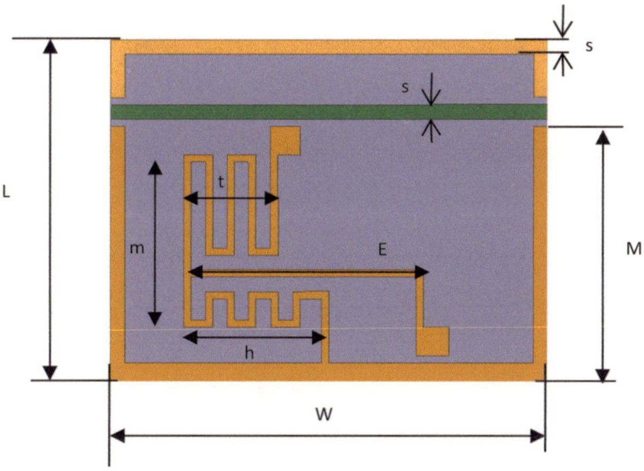

(a)

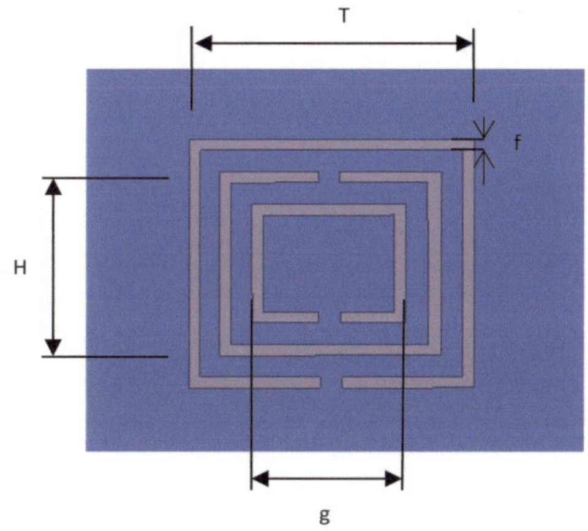

(b)

Figure 10.
(a) Top view, (b) bottom view of the electromagnetic sensor, CSSR.

L	W	H	E	M	T
15	15	6	8	8.25	7.5
h	s	t	m	f	g
6.75	0.5	4.5	7	0.25	4.5

Table 2.
Parameters of the proposed sensor in (mm).

Figure 12, there is a remarkable difference in higher volume fraction and almost none in the lower volume fraction. The liquid under test is presented in **Figure 13**, the liquid is supposed to be placed on the top of sensor in circular shape, and the input port and the output port are on the same side as shown in the figure.

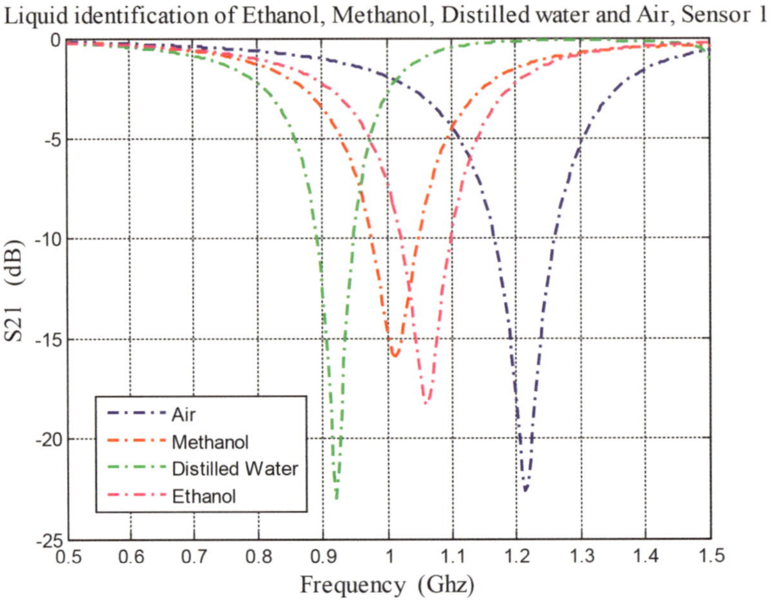

Figure 11.
Transmission coefficient with and without CSRR.

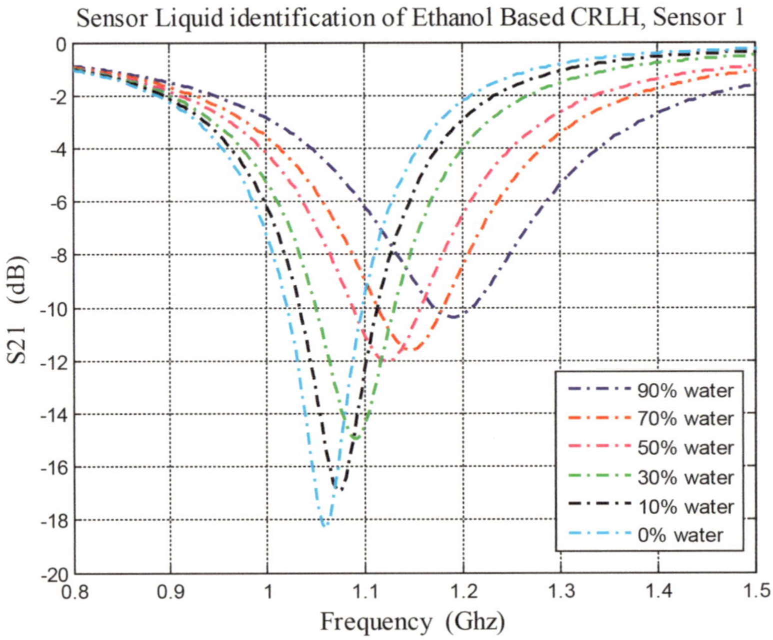

Figure 12.
Transmission coefficient of the volume fraction of ethanol.

According to the Debye relaxation model, the real part of the mixture permittivity can be expressed as follows [37–41]. Where ε_s is the low-frequency permittivity, ε_∞ is the high-frequency permittivity, τ is the relaxation time, and ω is the radiated frequency.

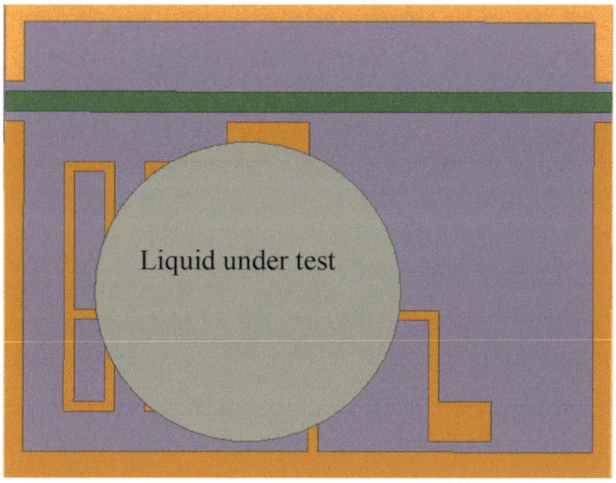

Figure 13.
Electromagnetic sensor loaded with liquid.

$$\varepsilon' = \varepsilon_{\infty} + \frac{\varepsilon_s - \varepsilon_{\infty}}{1 + \omega^2 \tau^2} \tag{6}$$

Due to the values' stability of the permittivity around 1.5 GHz, the dispersion is not taken into account. We can express the sensor sensitivity by:

$$S = \frac{f_o - f_1}{f_o d\varepsilon_r} \tag{7}$$

where f_o is the lower frequency of sensor (1.065 GHz) and f_1 is a upper frequency of sensor (1.195 GHz), and $d\varepsilon_r$ is the variation in the relative permittivity from volume fraction of 10–100%, which is considered of (79–9.5 = 69.5), from **Table 3**. The simulated sensitivity has a maximum value given as:

$$(1.195\text{–}1.065)/(1.195 \times 69.5) = 0.156\%$$

A sensitivity of 0.156%, which is acceptable compared with the literature in [29, 35, 37, 39, 40], given respectively as 0.11%, 0.27%, 0.54%, 0.26%, and 0.14%. The

	Ethanol	ϵ'	ϵ''
100%	0%	79.0	9.2
90%	10%	73.0	11.3
70%	30%	58.5	17.1
50%	50%	43%	17.1
30%	70%	29.0	16.0
10%	90%	15.5	12
0%	100%	9.5	8.1

Table 3.
References values of complex permittivity with mixture of volume fraction water-ethanol [23], at about 2 GHz.

sensitivity of the sensor could be improved by various implementations of metamaterial structures, even by adding electromagnetic band gap (EBG) structures at the top side, or defected ground at the bottom side of the sensor.

6. Conclusions

In this chapter, a great attention was given to the metamaterial structures, split ring resonators (SRRs), complementary split ring resonators (CSRRs), and complementary right-left hand (CRLH) structures, based on transmission lines technologies and zero-order resonators (ZORs). The study has included historical view on the metamaterials' appearance, development, and application. Physical properties of metamaterials have also been discussed, and different types of MTMs have been introduced, with MGN permeability negative, ENG negative dielectric, and DNG double-negative electric and magnetic components.

The first application of metamaterial was given to an ultrawide-band monopole antenna loaded with zero-order resonators (ZORs), the antenna could be used in WLAN, WiMax, and X band applications. The proposed antenna was designed and compared with three types of ZOR antenna of similar types. The computed results show that, when loading the monopole antennas with ZOR, the gain reaches 1.8 dBi, and the impedance bandwidth is improved by 148% compared with the original antenna. It was shown that the resonant frequency can be shifted toward lower or higher operating frequencies just by changing the inductor length or adjusting the capacitor value.

The second application is a very compact sensor based on CRLH resonator and CSRR, the sensor is designed for liquid mixture identification, the simulation is done on the commercial software HFSS, three types of liquid mixture are tested on this sensor, and these are ethanol, methanol, and distilled water. The obtained results have shown that the sensitivity of the sensor is very acceptable compared with the literature, and the dimension of the sensor is very compact as well as the testing surface of the sensor.

Finally, metamaterials have the ability of improving the performances of any microwave structures, and this can be done by increasing the gain or the impedance bandwidth of antennas or improving the sensitivity of electromagnetic sensors.

Author details

Mohamed Lashab[1*], Mounir Belattar[2], Sekkache Hocine[2] and Halfaya Ahmed[1]

1 Laboratory of Electronics and New Technology (LENT), University Larbi Ben M'Hidi, Oum El-Bouaghi, Algeria

2 University of 20 Aout, LARES Laboratory, Skikda, Algeria

*Address all correspondence to: lashabmoh@yahoo.fr

References

[1] Milligan TA. Modern Antenna Design. 2nd ed. New Jersey: Wiley & sons; 2005

[2] Balanis CA. Antenna Theory, Analysis and Design. 4th ed. New Jersey: Wiley & Sons; 2016

[3] Croswell WF, Bailey MC. In: Johnson RC, Jasik H, editors. Antenna Engineering Handbook. New York: McGraw-Hill; 1984

[4] Fujimoto K, James JR. Mobile Antenna Systems Handbook. 2nd ed. Boston: Artech House; 2001

[5] Tsoulos G. MIMO Systems Technology for Wireless Communication. New York: Taylor Francis; 2006

[6] Lashab M, Zebiri C-E, Djouablia L, Belattar M, Saleh A, Benabdelaziz F, et al. Characterization of horn antenna loaded with CLL unit cell. Microwave and Optical Technology Letter. 2018;**60** (8):1847-1856

[7] Hocine S, Lashab M, Ouchtati S. Design of sensor based on CRLH for liquid mixture application. In: ICEECA'19 4th International Conference on Electrical Engineering Control Application. Constantine Algeria; 2019

[8] Lashab M, Kosha J, Abdussalam FM, Belattar M, Djouablia L, Zebiri CE, et al. Design and Optimization of Electromagnetic Sensor Based On CRLH Resonator. Loughborough, UK; 2018

[9] Jan NA, Lashab M, Zebiri CE, Linda D, Abd-Alhameed RA, Benabdelaziz F. Compact CPW antenna loaded with CRLH-TL and EBG for multi-band and gain enhancement. In: Loughborough Antennas & Propagation Conference. Loughborough, UK: LAPC. 2016

[10] Ramsay J. Microondas antena de guía de ondas y técnicas antes de 1900. Actes de l'IRE (Abstracto). 1958;**46**(2):405

[11] Brown J. Artificial dielectrics. Progress in Dielectrics. 1960;**2**:195-225

[12] Kock WE. Metal-lens antennas. Proceedings of IRE. 1946;**34**:828-836

[13] Shelby R et al. Experimental verification of a negative index of refraction. Science. 2001;**292**:77-79

[14] Lamb H. On group-speed. Proceedings of London Mathematical Society. 1904;**1**:473-479

[15] V. Veselago. "The electrodynamics of substances with simultaneously negative values of ε and μ," Soviet Physics Uspekhi, vol. 10, no. 4, pp. 509–514, Jan., Feb. 1968

[16] Pendry JB. Negative refraction makes a perfect lens. Physical Review Letters. 2000;**85**(18):3966-3969

[17] Caloz C, Itoh T. Novel microwave devices and structures based on the transmission line approach of meta-materials. In: IEEE-MTT Int'l Symp. Philadelphia, PA; 2003. pp. 195-198

[18] A. Sanada, C. Caloz, and T. Itoh. "Characteristics of the composite right/left-handed transmission lines," IEEE Microwave Wireless Components Letters, vol. 14, no. 2, pp. 68–70, February 2004

[19] Caloz C, Sanada A, Itoh T. A novel composite right/left-handed coupled-line directional coupler with arbitrary

coupling level and broad bandwidth. IEEE Transactions on Microwave Theory Technology. 2004;**52**(3):980-992

[20] Engheta N, Ziolkowski RW. Metamaterials, Physics and Engineering Explorations. Canada: Wiley & Sons; 2006

[21] D. R. Smith, W. J. Padilla, D. C. Vier, S. C. Nemat-Nasser, and S. Schultz. "Composite medium with simultaneously negative permeability and permittivity, " Physical Review Letters, vol. 84, no. 18, pp. 4184–4187, May 2000

[22] Smith DR, Schurig D. Electromagnetic wave propagation in media with indefinite permittivity and permeability tensors. Physical Review Letters. 2003;**90**:077405

[23] Bao J-Z, Swicord ML, Davis CC. Microwave dielectric characterization of binary mixtures of water, methanol, and ethanol. Journal of Chemical Physics. 1996;**104**(12):22

[24] Lashab M et al. Electrically Small Planar Antennas Based on Metamaterial: Antenna Fundamentals for Legacy Mobile Applications and Beyond. Springer International Publishing AG; 2018. pp. 71-98. DOI: 10.1007/978-3-319-63967-3_4

[25] Sabah C, Uckun S. Multilayer system of Lorentz-Drude type Metamaterials with dielectric slabs and its application to electromagnetic Filters. Progress in Electromagnetics Research. 2009;**91**: 349-364

[26] Smith DR, Shultz S, Markos P, Soukoulis CM. Determination of effective permittivity and permeability of Metamaterials from reflection and transmission coefficients. Physical Review B. 2002;**65**:1951041-1951045

[27] Chen X, Grzegorczyk TM, Wu BI, Pacheco J, Kong JA. Robust method to retrieve the constitutive effective parameters of metamaterials. Physical Review E. 2004;**70**:016608

[28] Lubkowski G, Schuhmann R, Weiland T. Extraction of effective metamaterial parameters by parameter fitting of dispersive models. Microwave Optic Technology Letters. 2007;**49**(2): 285-288

[29] Saeed K, Pollard RD, Hunter IC. Substrate integrated waveguide cavity resonators for complex permittivity characterization of materials. IEEE Transaction on Microwave Theory and Techniques. 2008;**56**(10):2340-2347

[30] Chi Y-J, Chen F-C. Compact CPW-based zeroth-order Resonant Antenna with interleaving CRLHD unit cells. Progress in Electromagnetics Research C. 2013;**40**:119-130

[31] Wang G, Feng Q. A novel coplanar waveguide feed zeroth-order resonant antenna with resonant ring. IEEE Antennas and Wireless Propagation Letters. 2014;**13**:774-777

[32] Lashab M, Jan NA, Zebiri C. The I shape Antenna Loaded with ZOR For WLAN and WiMax Application. Loughborough; 2015

[33] Chaozhu Zhang, Jing Zhang and Lin Li, "Triple band-notched UWB antenna based on SIR-DGS and fork-shaped stubs," Electronics Letters 16th January 2014 Vol. 50 No. 2 pp. 67–69

[34] Withawat Withayachumnankul, Kata Jaruwongrungsee, Christophe Fumeaux and Derek Abbott , "Metamaterial-inspired multichannel thin-film sensor," IEEE Sensors Journal, Vol. 12, No. 5, May 2012

[35] Ebrahimi A, Withayachumnankul W, Al-Sarawi S, Abbott D. High-sensitivity metamaterial-inspired sensor for microfluidic dielectric characterization. IEEE Sensors Journal. 2014;**14**(5)

[36] Horestani AK, Naqui J, Abbott D, Fumeaux C, Martín F. Two-dimensional displacement and Alignment sensor based on reflection Coefficients of open microstrip lines loaded with split ring resonators. Electronics Letters. 2014; **50**(8):620-622

[37] Benkhaoua L, Benhabiles MT, Riabi ML. Miniaturized quasi-lumped resonator for dielectric characterization of liquid mixtures. IEEE Journal & Magazine. 2016;**16**(6):1603-1610

[38] Albishi AM, Ramahi OM. Microwaves-based high sensitivity sensors for crack detection in metallic materials. IEEE Transaction on Microwave Theory and Techniques. 2017;**65**(5):1864-1872

[39] Chretiennot T, Dubuc D, Grenier K. "A microwave and microfluidic planar resonator for efficient and accurate complex permittivity characterization of aqueous solutions," Microwave Theory and Techniques, IEEE Transactions on. Feb 2013;61(2):972,978

[40] Gennarelli G, Romeo S, Scarfi MR, Soldovieri F. A microwave resonant sensor for concentration measurements of liquid solutions. Sensors Journal IEEE. 2013;**13**(5):1857

[41] Hocine S, Mohamed Lashab M, Belattar SO, See CH, Yim-Fun H, Abd-Alhameed RA. Microwave sensor for liquid mixture identification based on composite right left hand-zero-order resonator for sensitivity improvement. International Journal of RF and Micro Computer Aided Engineering. Wiley; 2022;**32**(11):1456-1465

Section 3

Metamaterials in Absorbers

Chapter 5

An Application-Based Study on Electromagnetic Absorber Using Metamaterial

Alkesh Agrawal

Abstract

In recent years, metamaterials (MMs) have attracted researchers due to their geometrical and structural uniqueness that make these materials to absorb, block, and enhance electromagnetic (EM) waves, which is not possible with conventional materials found in nature. These artificially engineered materials derive the EM properties (effective values of permittivity $\tilde{\varepsilon}_{\text{eff}}$ and permeability $\tilde{\mu}_{\text{eff}}$ less than zero) from the shape, size, orientation, and periodicity of unit cells rather inheriting those from material composition. The study on MMs has been diversified from the radio frequency range to the optical frequency range, with potential applications in realization of novel devices such as perfect lenses, EM, and MM based microwave patch antennas. For the past few years, the concept of MMs has been widely used to develop and design metamaterial perfect absorbers (MPAs). The proposed chapter mainly focuses on the classification of materials on the basis of permittivity and permeability; MPAs; applications of MPAs; experimental demonstrations of first single-band MPAs in microwave, THz, mid-IR and near IR regimes; conditions for complete absorption of EM waves; MPA as perfectly matched layer (PML); attenuation mechanism of EM waves inside the MPA; calculation of MM parameters; measurement and testing process, followed by a case study on multi-band MPA.

Keywords: absorptance, metamaterial, matched impedance, permittivity, permeability

1. Introduction

The electromagnetic (EM) properties of materials are characterized on the basis of negative or positive values of permittivity $\tilde{\varepsilon}$ or permeability $\tilde{\mu}$. Most of the materials that are found in nature have positive values of permittivity $\tilde{\varepsilon}$ and permeability $\tilde{\mu}$, and hence, they are termed as "double-positive" (DPS) materials [1, 2]. Materials exhibiting negative values of permittivity or permeability are termed as "single-negative" (SNG) and these materials are further classified into two sub-categories "epsilon-negative" (ENG) and "mu-negative" (MNG) [1, 2]. If these quantities, permittivity $\tilde{\varepsilon}$ and permeability $\tilde{\mu}$ are negative, they are called "double-negative" (DNG) [1, 2]. The energy flux carried by the EM wave is determined by the Poynting vector $\overline{S}$, given by Eq. (1) [1].

$$\overline{S} = \frac{c}{4\pi}\left(\overline{E} \times \overline{H}\right) \tag{1}$$

For DPS materials the vectors, $\overline{E}$ (electric field vector), $\overline{H}$ (magnetic field vector), and $\overline{k}$ (propagation vector) form a right handed set of vectors, where $\overline{S}$ (Poynting vector) and $\overline{k}$ are co-directional. The DPS materials are also termed as "right-handed materials" (RHMs). For DNG materials, the same set of vectors forms left-handed set of vectors, but vectors $\overline{S}$ and $\overline{k}$ are in opposite directions. The DNG materials are also termed as "left- handed materials" (LHMs). The LHMs are also termed as substances with negative group velocity which means that the phase velocity is in opposite direction to that of energy flux.

The electric and magnetic fields are associated with each other, however the permittivity $\tilde{\varepsilon}$ and the permeability $\tilde{\mu}$ control electric and magnetic fields in a medium. Negative permeability results in orientation of magnetic dipoles in reverse direction so as to oppose the external applied magnetic field (diamagnetic materials, anti-ferro magnetic materials), similarly the concept of negative permittivity results in the polarization of electric dipoles in reverse direction. Thus, by making both the permittivity $\tilde{\varepsilon}$ and the permeability $\tilde{\mu}$ negative, the direction of flow of energy inside the material is made to change in contradiction to the conventional direction of flow of energy. The natural materials do not possess simultaneous negative values of permittivity $\tilde{\varepsilon}$ and permeability $\tilde{\mu}$, hence they must be created artificially (**Figure 1**).

MMs are artificially engineered new class of materials that are not found in nature. These artificial materials possess unusual EM properties with values of effective permittivity $\tilde{\varepsilon}_{\text{eff}}$ and effective permeability $\tilde{\mu}_{\text{eff}}$ less than zero [3]. The geometry, the periodic arrangement and the orientation of the structural units give rise to unusual EM properties. These unusual EM properties are not inherited from the material composition. So, MMs can be defined as periodic arrangement or irregular periodic

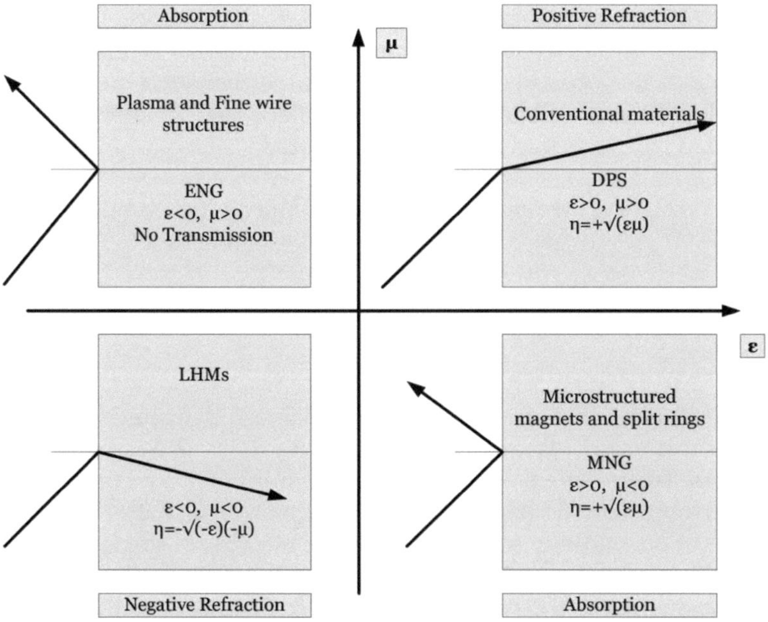

Figure 1.
Classification of known materials on the basis of $\tilde{\varepsilon}$ and $\tilde{\mu}$ with the realization of metamaterials (MMs) [1].

Figure 2.
Experimental demonstration of composite medium by Shelby et al. [2].

arrangement of sub-wavelength metallic structures which derive EM material properties from their geometry rather than inheriting them directly from the material composition or band structure.

In 1968 Veselago [1] gave theoretical concept of these new class of materials that can possess unusual properties like negative index of refraction, opposite phase and group velocity, reverse Doppler shift, reverse Cerenkov radiations. The absence of the natural occurrence of materials with these properties led to neglect of the subject until 1996 when Pendry et al. [4] explained the design and behavior of artificial material termed as metallic microstructures comprising of periodic structure of infinite wires arranged in cubic lattice and exhibiting negative permittivity $\tilde{\varepsilon}$ and split ring resonator (SRR) structures exhibiting negative permeability $\tilde{\mu}$. In 1999 Smith et al. [3] demonstrated a composite medium exhibiting simultaneous negative values of effective permeability $\tilde{\mu}_{\text{eff}}$ and permittivity $\tilde{\varepsilon}_{\text{eff}}$ comprised of periodic array of SRRs and continuous micro-structured wires in microwave region, forming left-handed medium (**Figure 2**). With this demonstration the concept of materials with negative values of $\tilde{\varepsilon}_{\text{eff}}$ and $\tilde{\mu}_{\text{eff}}$ given by Veselago way back in 1968 came into existence and a whole new field of research came into limelight to be known as 'Metamaterials'. Word 'Metamaterial' is derived from Greek word 'meta' which means 'beyond'. The term metamaterial was first introduced by Walser [5].

The geometrical and structural uniqueness of MM makes these materials capable to bend [6, 7], absorb [8, 9], block [10, 11], and enhance [12] EM waves which is not possible with conventional materials found in nature [1]. In recent years the study of MMs has been diversified from radio frequency range to optical frequency range.

The MMs find applications to realize perfect-lens [6], super-lens [7], EM cloak [10, 11], EM concentrators [12], EM band gap based microwave circuit design [13], multi-band MNG resonators [14], MM-based and MM inspired efficient, electrically small antennas [15], MM-based patch and leaky-wave antennas [16], MM-based perfect absorbers [8, 9, 17].

2. EM wave absorber

An EM wave absorber is a device which absorbs all the EM radiation incident on it under perfectly matched conditions. American engineer Salisbury and another scientist J. Jaunmann [18] invented EM absorbers independently to improve the performance of radar and to provide stealth technology. Salisbury [18] developed the

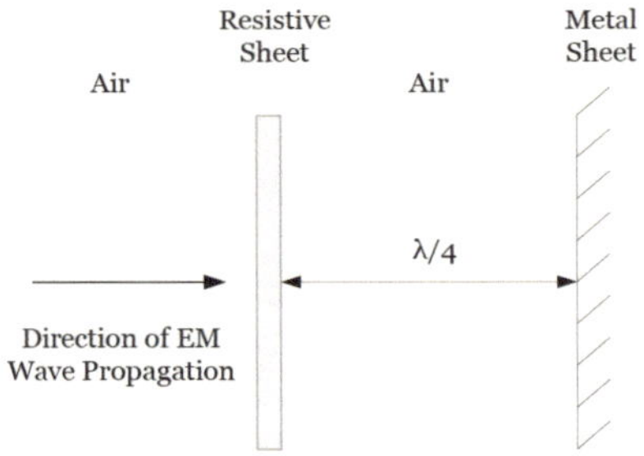

Figure 3.
Salisbury screen [18].

Salisbury screen. The basic structure of EM wave absorber consists of a resistive sheet and metal laminated ground plane both separated by some lossless dielectric of thickness ($\lambda_o/4$) as shown in **Figure 3**. When EM waves are incident on the Salisbury screen, the waves pass through the resistive sheet without any reflections and the EM waves get reflected by the continuous metallic surface. The waves travel back $\lambda/2$ distance with phase change by 180° and result in destructive interference with incident EM waves.

The performance of Salisbury screen is limited firstly because of the bulky size. The thickness increases for broad-band absorption due to cascading of dielectric lossy layers. To satisfy the complete destructive interference the thickness of dielectric layer must be $\left(\frac{\lambda/4}{\eta}\right)$. If wavelength ($\lambda$) of EM wave is large then the dielectric layer has to be very thick. Secondly, the performance of Salisbury screen is limited to microwave frequencies due to impedance matching conditions.

3. Metamaterial perfect absorber

Conventional metamaterial perfect absorber (MPA) is a three-layered structure [8]. It consists of top layer that is periodic array of unit cells, constituting MM high impedance surface, a middle layer that is lossy or lossless dielectric layer, and a bottom layer that is metal laminated ground plane. **Figure 4** shows the layered structure of

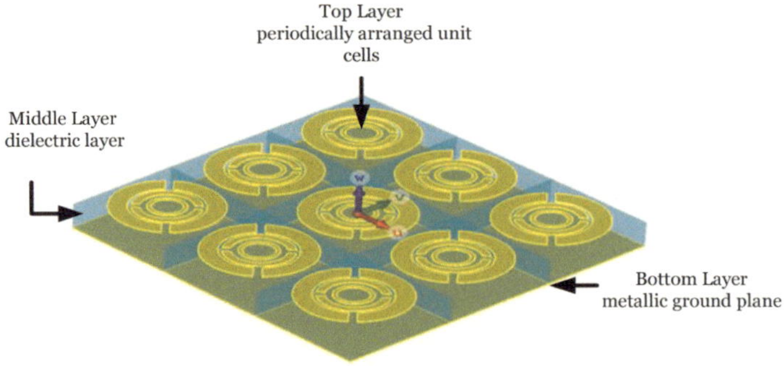

Figure 4.
Layered structure of MPA [17].

MPA. The shape, size, periodicity, and orientation of the unit cells are optimized to match the normalized impedance of the top layer of MPA with that of the impedance of free space. Under normalized impedance matched conditions, the incident EM waves at certain frequency bands propagates through the high impedance layer without any reflections. The continuous metal laminated bottom layer completely blocks the EM waves and reflects back the EM waves.

The dielectric layer provides space to the incident EM waves to stay and get absorbed inside the material. The dielectric layer should have high permittivity as it results in reduction of thickness maintaining the optical path. The dielectric layer should also have high loss tangent (tan δ) value. The MPA must offer high values of imaginary part of effective permittivity and effective permeability to attenuate the EM waves inside the MPA.

The complete absorption of the incident microwave radiation from free space on a microwave absorber surface requires the fulfillment of the following conditions:

a. The complete transfer of the incident microwave radiation into the surface of the microwave absorber which can be achieved by perfect impedance matching of the free space and the front surface of the microwave absorber known as perfectly matched condition.

b. The transferred microwave radiation into the microwave absorber should be completely absorbed within the microwave absorber which can be achieved by high attenuation constant for the incident microwave radiation inside the microwave absorber.

3.1 Metamaterial perfect absorber as perfectly matched layer

A conventional MPA structure consisted of three-layered structure, the uppermost high impedance layer described by effective impedance $\tilde{Z}_{\text{eff}}(f) = [\tilde{\mu}_{\text{eff}}(f)/\,\tilde{\varepsilon}_{\text{eff}}(f)]^{1/2}$, the middle dielectric layer described by effective permeability $\tilde{\mu}_{\text{eff}}(f)$ and effective permittivity $\tilde{\varepsilon}_{\text{eff}}(f)$ and lowermost continuous metal laminated ground layer. Berenger in 1994 [19], gave the idea of artificial absorbing layer which can be frequency and angle of incident independent described as perfectly matched layer (PML). The PML has the characteristic of no reflection at the interface and maximum absorption inside the medium. To achieve this characteristic, the PML has its effective impedance matched with that of free space. i.e. $\tilde{Z}_{\text{eff}}(f) = 1$.

According to the Fresnel's equations [20], the reflectance (R) and reflection coefficient (γ) of the interface separating free space (with impedance, permeability, permittivity and, refractive index: Z_o, μ_o, ε_o, η_o, respectively) and medium (with impedance, permeability, permittivity and, refractive index: $\tilde{Z}$, $\tilde{\mu}_r$, $\tilde{\varepsilon}_r$, $\tilde{\eta}$, respectively shown in **Figure 5**) for TE and TM polarized wave is given by Eq. (2) and Eq. (3), respectively [20].

The impedance $\tilde{Z}$, permeability $\tilde{\mu}$, permittivity $\tilde{\varepsilon}$, and refractive index $\tilde{\eta}$ of the media are complex quantities [21] defined by Eq. (4), Eq. (5), Eq. (6), and Eq. (7), respectively, and the effective impedance $\tilde{Z}_{\text{eff}}$, permeability $\tilde{\mu}_{\text{eff}}$, permittivity $\tilde{\varepsilon}_{\text{eff}}$, and refractive index $\tilde{\eta}_{\text{eff}}$ of the media are defined by Eq. (8), Eq. (9), Eq. (10), and Eq. (11), respectively.

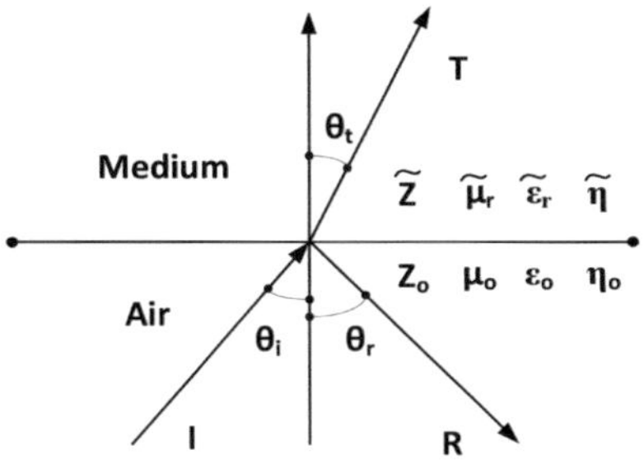

Figure 5.
Incident (I), transmitted (T) and reflected (R) EM wave.

$$R_{\text{TE}} = |\gamma_{\text{TE}}|^2 = \left|\frac{\text{Cos}\,\theta_i - \tilde{\mu}_r^{-1}\sqrt{\tilde{\eta}^2 - \text{Sin}\theta_i^{\,2}}}{\text{Cos}\,\theta_i + \tilde{\mu}_r^{-1}\sqrt{\tilde{\eta}^2 - \text{Sin}\theta_i^{\,2}}}\right|^2 \tag{2}$$

$$R_{TM} = |\gamma_{TM}|^2 = \left|\frac{\tilde{\varepsilon}_r\text{Cos}\,\theta_i - \sqrt{\tilde{\eta}^2 - \text{Sin}\theta_i^{\,2}}}{\tilde{\varepsilon}_r\text{Cos}\,\theta_i + \sqrt{\tilde{\eta}^2 - \text{Sin}\theta_i^{\,2}}}\right|^2 \tag{3}$$

where θ_i is the angle of incidence and $\tilde{\eta} = \sqrt{\tilde{\mu}_r \tilde{\varepsilon}_r}$ is the refractive index of the dielectric medium.

$$\tilde{Z} = Z'(f) + jZ''(f) \tag{4}$$

$$\tilde{\mu} = \mu'(f) + j\mu''(f) \tag{5}$$

$$\tilde{\varepsilon} = \varepsilon'(f) + j\varepsilon''(f) \tag{6}$$

$$\tilde{\eta} = \eta'(f) + j\eta''(f) \tag{7}$$

$$\tilde{Z}_{\text{eff}} = \frac{\tilde{Z}}{Z_o} \tag{8}$$

$$\tilde{\mu}_{\text{eff}} = \frac{\tilde{\mu}}{\mu_o} \tag{9}$$

$$\tilde{\varepsilon}_{\text{eff}} = \frac{\tilde{\varepsilon}}{\varepsilon_o} \tag{10}$$

$$\tilde{\eta}_{\text{eff}} = \frac{\tilde{\eta}}{\eta_o} \tag{11}$$

where $Z'(f)$, $\mu'(f)$, $\varepsilon'(f)$ and $\eta'(f)$ are the real part and $Z''(f)$, $\mu''(f)$, $\varepsilon''(f)$ and $\eta''(f)$ are the imaginary part of impedance, permeability, permittivity, and refractive index of the medium, respectively. For normal angle of incidence θ_i = 0 degree, the Eq. (2) and Eq. (3) reduced to Eq. (12).

$$R = \left|\frac{\tilde{Z} - Z_o}{\tilde{Z} + Z_o}\right|^2 \tag{12}$$

where $\tilde{Z} = \sqrt{\frac{\tilde{\mu}}{\tilde{\varepsilon}}}$ and that of free space is $Z_o = \sqrt{\frac{\mu_o}{\varepsilon_o}}$.

The absorptance, A = $[1 - R - T]$, where R is the reflectance and T is the transmittance. Due to the presence of the bottom ground layer, T is reduced to zero and absorptance $A = [1 - R]$. For perfect absorption, i.e., $A = 1$, the reflectance R should be zero which can be achieved if, $\tilde{Z} = Z_o$. This condition is referred to as perfect matched condition [8, 9].

Attainment of perfectly matched condition so as to have minimum reflection from the material and maximum dissipation of EM waves inside the material led to a new field of research identified as MPAs. To achieve the perfect absorption of incident EM wave radiation researchers explored MM design concept to achieve a material having absorptance of 'Unity', i.e., MPA. After the first experimental demonstration of MPA [8], advanced research has been carried out in the field of MPA from radio spectrum to optical spectrum at microwaves, mm wave, THz, infrared, and optical range. The advancement and improvement in designing MPAs to maximize the absorptance over a wide frequency band contributes to narrow-band MPAs, dual-band and multi-band MPAs, broad-band MPAs, polarization insensitive MPAs, and wide angle MPAs. MPAs find potential applications in reducing radar cross section (RCS) [22], in improving antenna radiation pattern [16], and in reducing electromagnetic interference (EMI) [21]. Future applications include the use of MPA as selective thermal emitters [23] and wavelength sensitive sensors [24].

3.2 Metamaterial parameters

Under impedance matched conditions, the MPA structure is considered to be a single homogeneous layer with dispersive effective permittivity $\tilde{\varepsilon}_{\text{eff}}(f)$ and effective permeability $\tilde{\mu}_{\text{eff}}(f)$. To study the absorptance of incident EM waves through frequency selective top layer of MPA, the effective impedance $\tilde{Z}_{\text{eff}}(f)$, the effective permittivity $\tilde{\varepsilon}_{\text{eff}}(f)$, the effective permeability $\tilde{\mu}_{\text{eff}}(f)$, and the effective refractive index $\tilde{\eta}_{\text{eff}}(f)$ is calculated.

According to Nicolson-Ross-Weir (NRW) method [25], the effective impedance $\tilde{Z}_{\text{eff}}(f)$ is calculated using Eq. (17) given as:

$$\tilde{Z}_{\text{eff}}(f) = \frac{1 + S_{11}(f)}{1 - S_{11}(f)} \tag{13}$$

The effective permittivity $\tilde{\varepsilon}_{\text{eff}}(f)$ is calculated using Eq. (18) given as:

$$\tilde{\varepsilon}_{\text{eff}}(f) = \frac{2}{\mathrm{j}\, k_o d}\left(\frac{1 - S_{11}(f) - S_{21}(f)}{1 + S_{11}(f) + S_{21}(f)}\right) \tag{14}$$

The effective permeability $\tilde{\mu}_{\text{eff}}(f)$ is calculated using Eq. (19) given as:

$$\tilde{\mu}_{\text{eff}}(f) = \frac{2}{\mathrm{j}\, k_o d}\left(\frac{1 + S_{11}(f) - S_{21}(f)}{1 - S_{11}(f) + S_{21}(f)}\right) \tag{15}$$

The effective refractive index $\tilde{\eta}_{\text{eff}}(f)$ is calculated using Eq. (20) given as:

$$\tilde{\eta}_{\text{eff}}(f) = \sqrt{\tilde{\varepsilon}_{\text{eff}}(f) * \tilde{\mu}_{\text{eff}}(f)} \tag{16}$$

In an MPA, under perfectly matched conditions, at the maximum absorptance frequency (f_o), the real and imaginary parts of effective impedance of MPA matches with that of free space. $\tilde{Z}_{\text{eff}}(f)\big|_{f=f_o} = Z'_{\text{eff}}(f_o) + i\,Z''_{\text{eff}}(f_o)$ = 1. It implies that the normalized real value of $\tilde{Z}_{\text{eff}}(f)$ is equal to 1 and the imaginary value of $\tilde{Z}_{\text{eff}}(f)$ is equal to zero.

$$\text{Re}\,[\tilde{\varepsilon}_{\text{eff}}(f)]_{f=f_o} = Re\,[\tilde{\mu}_{\text{eff}}(f)]_{f=f_o} \text{ and } \text{Im}[\tilde{\varepsilon}_{\text{eff}}(f)]_{f=f_o} = \text{Im}[\tilde{\mu}_{\text{eff}}(f)]_{f=f_o}$$

In case of MM behavior, the real part of $\tilde{\varepsilon}_{\text{eff}}(f)$ and $\tilde{\mu}_{\text{eff}}(f)$ shows the Lorentzian oscillation at or within the vicinity of maximum absorptance frequency band. The imaginary part of $\tilde{\varepsilon}_{\text{eff}}(f)$ and $\tilde{\mu}_{\text{eff}}(f)$ attains appreciable high values as compared to real part. Under perfectly matched condition $\tilde{Z}_{\text{eff}}(f)$ = 1, the point at which the real values of $\tilde{\varepsilon}_{\text{eff}}(f)$ and $\tilde{\mu}_{\text{eff}}(f)$ crosses $\tilde{Z}_{\text{eff}}(f)$=1, the imaginary values of $\tilde{\varepsilon}_{\text{eff}}(f)$ and $\tilde{\mu}_{\text{eff}}(f)$ attains appreciable high values as shown in **Figure 6a** and **b**.

3.3 Measurement and testing of metamaterial absorber

The measurement and testing setup consisted of anechoic chamber, transmitting, and receiving horn antennas, and vector network analyzer (VNA) as shown **Figure 7**. The measurement process is calibrated in order to reduce the measurement errors. The measurement process is carried out in two different steps as described below.

a. In first step the MPA structure is placed in front of transmitting and receiving horn antennas with copper laminated ground plane of MPA structure facing the horn antennas in an anechoic chamber as shown in **Figure 7**. The transmitting horn antenna is connected to port 1 of VNA and the receiving horn antenna is connected to the port 2 of the VNA as shown in **Figure 7**. The transmitting horn antenna transmits 1 mW EM radiation towards the copper laminated ground plane of MPA. The reflected radiation from the copper laminated ground plane of MPA is received by the receiving horn antenna. The VNA measures the reflection coefficients corresponding to the reflected power.

b. In second step the MPA structure is rotated by 180° such that the unit cells structured side of MPA faces the transmitting and receiving horn antennas. The reflected radiation from the structure side of MPA is received by the receiving horn antenna. The VNA again measures the reflection coefficients corresponding to the reflected power. The difference of the two measurements gives the calibrated measured values.

The two-step measuring process is used to measure absorptance for normal incidence, for variation in angle of incidence θ, and variation in E-field and H-field

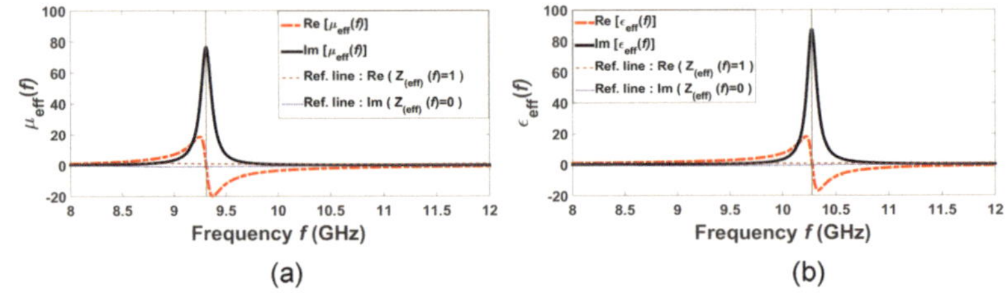

Figure 6.
Real and imaginary values of (a) $\tilde{\mu}_{\text{eff}}(f)$ (b) $\tilde{\varepsilon}_{\text{eff}}(f)$ at or within maximum absorptance frequency band [17].

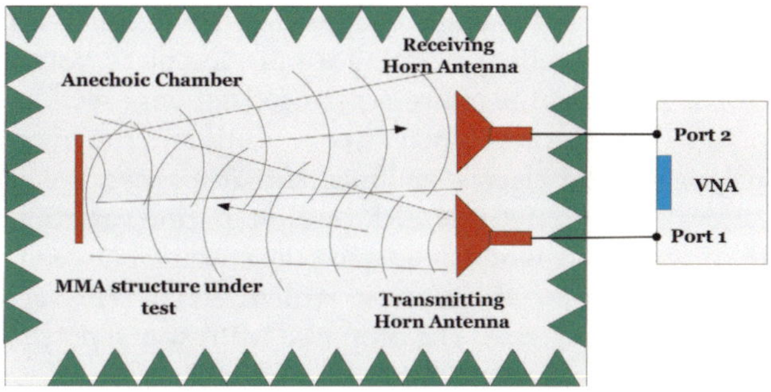

Figure 7.
Block diagram of the experimental set up.

orientations Φ of the incident EM waves. For taking measurements of reflection coefficients S_{11} at different angles of incident of EM waves, the MPA structure sample placed on a circular scale is rotated along the y-axis by the desired angle and measurements were recorded in two steps as described. Similarly for taking measurements of reflection coefficients S_{11} at different angles of E-field and H-field of EM waves, the MPA structure sample is rozztated along the z-axis by the desired angle.

4. First reported single-band metamaterial perfect absorbers

The first narrow single-band MPA was experimentally demonstrated in 2008 by Landy et al. [8]. The unit cell consisted of electric ring resonator (ERR) over micro-structured cut wired section separated by dielectric layer as shown in **Figure 8**. The dielectric layer was of FR-4 substrate of 0.2 mm thickness. The top-layered ERR and bottom-layered micro-structured cut wired section was etched out of 17 μm thick copper layers shown in **Figure 8**. The MMA structure consisted of two-dimensional arrays of unit cells with separation of 0.72 mm. The MPA was fabricated using photosensitized method. The results simulated in finite difference time domain (FDTD) solver, computer simulation technology (CST) microwave studio showed absorptance of 96% at 11.48 GHz whereas experimentally achieved absorptance was 88% at 11.5 GHz. The difference in the simulated and experimental values was mainly explained due to fabrication errors. The top layer comprising of ERR and the bottom layer comprising of

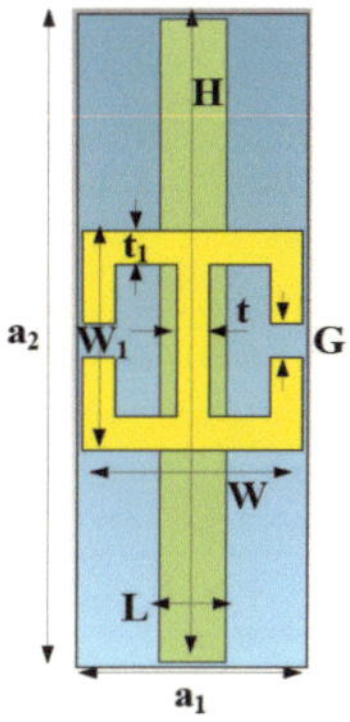

Figure 8.
Top view of narrow band MPA experimentally demonstrated in 2008 by Landy et al. [8].

micro-structured cut wire section strongly coupled the incident EM waves at the resonance frequency 11.48 GHz thereby generating strong electric response. The anti-parallel currents induced in the ERR and the micro-structured cut wire section was responsible for magnetic coupling. Thus, the combined effect of ERR and the micro-structured cut wire section was responsible for electric and magnetic responses. It was explained by the authors that the variation in geometry of ERR resulted in fine variation in absorptance frequency strength of resonance whereas varying the spacing between the two metallic structures ERR top layer and the bottom micro-structured cut wire section layer resulted in modification of magnetic response. The proposed MPA was reported to have potential application in devices such as bolometers.

After the first experimental demonstration of narrow single-band MPA in microwave region, the same work was extended in THz region [9] with MPA dimensions in μm and replacing FR-4 substrate layer with polyamide dielectric substrate layer as indicated in **Table 1**. The simulated results in CST microwave studio showed absorptance of 98% at 1.12 THz as compared to experimental results with 70% absorptance at 1.3 THz with potential application as low thermal mass absorber with thermal sensing.

The MPA structure was fabricated using surface micro-machining process with the deposition of 200 nm thick cut wire layer of Au/Ti ground plane on GaAs wafer followed by deposition of 8 μm thick layer of polyamide as dielectric substrate and on the top a 200 nm-thick Au/Ti layer as ERR.

The first single-band MPA in mid-IR regime was experimentally demonstrated in 2010 by Liu et al. [24]. The 140 × 140 μm MPA structure was fabricated with the deposition of 100 nm thick gold layer ground plane on silicon substrate followed by 185 nm thick layer of Al_2O_3 as dielectric substrate and on the top a 100 nm thick layer of gold in the cross wired shape (plus shape) as shown in **Figure 9a** and **b**. The simulated and measured absorptance was 97% at 6 μm and the absorptance band has

Type of absorber →	Microwave absorber (dimensions in mm)	THz absorber (dimensions in μm)
Resonance frequency	11.48 GHz	1.3 THz
Absorptance	88%	70%
a_1	4.2	34
a_2	12	50
t	0.6	3
t_1	0.6	3
G	0.6	3
W	4	30
W_1	4	30
H	1.7	4
L	11.8	48
Type of substrate	FR-4	Polyamide
Thickness of ERR and cut wire section	0.017	0.2
Thickness of substrate	0.2	8

Table 1.
Dimensional comparison of first experimentally demonstrated MPA in μW and THz regime.

FWHM of 1 μm. The proposed MPA has potential application in hyper-spectral imaging cameras.

The plus-shaped cross wired structure acted as ERR and it was responsible for coupling of E-field and the combination of ERR and ground plane for coupling of H-field from the incident EM radiation.

With the impedance of MPA matched to that of free space and the width of ground plane exceeding the penetration depth in mid-IR led to maximum EM radiation absorption with zero transmission through MPA.

The first experimental demonstrations for MPAs in NIR regime were reported by Hao et al. [26] and Liu et al. [27] separately in 2010. The MPA structure design by Liu et al. consisted of top layer of gold disks of 352 nm in diameter and 20 nm in height and gold lamination ground plane of 200 nm separated by MgF_2 dielectric of 30 nm thickness as shown in **Figure 10b**. The entire MPA structure was grown over glass substrate. The MPA structure was measured to be 99% polarization insensitive and to have wide angular absorptance at 1.6 μm with application as plasmonic sensor for refractive index sensing. The MPA structure design by Hao et al. consisted of top layer of rectangular patches of gold of 170 nm by 170 nm by 40 nm and gold lamination ground plane of 310 × 310 × 50 nm separated by Al_2O_3 dielectric of 10 nm thickness as shown in **Figure 10a**. The simulated absorptance was found to be greater than 97% at 1.58 and 1.95 μm for TM and TE radiation modes respectively. The ultra-thin MPA structure was measured to have 88% wide angular absorptance at 1.58 μm for TM mode radiations and 83% absorptance at 1.95 μm for TE mode radiations. The proposed MPA has micro technological applications including micro-bolometers, photo detectors, coherent thermal emitters, and solar cells. Anti-parallel surface currents induced in gold disks (or gold patches) and the ground plane gold layer resulting in magnetic resonance where the magnetic moment due to circulating currents strongly interact with magnetic field

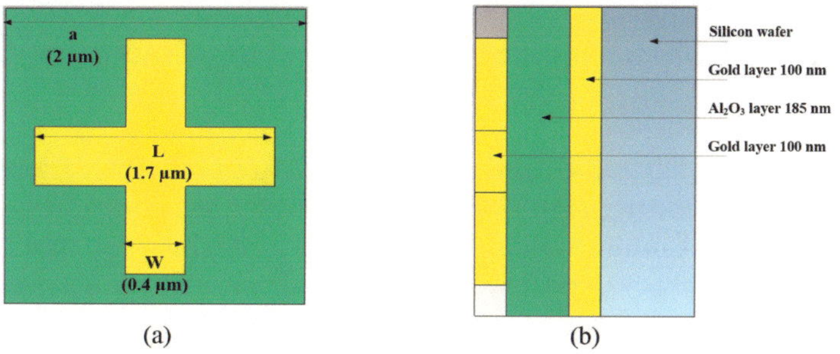

Figure 9.
Schematic of first single- band MPA in mid-IR regime experimentally demonstrated in 2010 by Liu et al. [24] (a) Top view. (b) Side view.

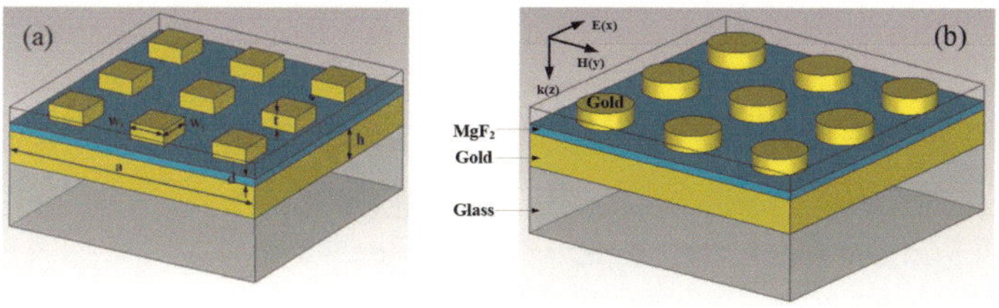

Figure 10.
(a) Schematic of first single-band MPA in NIR regime experimentally demonstrated in 2010 by Hao et al. [26].
(b) Schematic of first single-band MPA in NIR regime experimentally demonstrated in 2010 by Liu et al. [27].

Reference	Unit cell design	Frequency/ wavelength		Absorption bandwidth	Absorptance	
		Sim.	Exp.		Sim.	Exp.
[8]	Electric ring resonator (ERR) with back-to-back connected two split rings and a micro-structured split wire.	11.4 GHz	11.5 GHz	—	96%	88%
[9]	Electric ring resonator (ERR) with back-to-back connected two split rings and a micro-structured split wire. The dimensions of unit cell were in μm.	1.12 THz	1.3 THz	—	98%	70%
[24]	ERR as cross ring resonator (plus shape).	—	6 μm	FWHM :1 μm	—	97%
[26]	Rectangular patches of gold with gold laminated ground plane separated by Al_2O_3 dielectric.	1.58 μm	1.58 μm	—	97%	88%
[27]	Gold disks with gold laminated ground plane separated by MgF_2 dielectric.	—	1.6 μm	—	—	99%

Table 2.
Experimental demonstrations of first single-band MPAs in microwave, THz, mid-IR and near IR regime.

of incident EM radiation. At resonance the strong localized EM field enhanced between the two gold layers is confined in the dielectric layer and led to minimum reflectance and maximum absorptance of incident EM radiations (**Table 2**).

5. Case study: multi-band metamaterial absorber-simulation, fabrication, testing

The multi-band metamaterial absorbers (MMAs) are practically more useful when they are capable of exhibiting high absorptance at many resonance frequencies. The potential applications of multi-band microwave absorbers increase manifolds if they are insensitive to state of polarization and wide angular incident microwaves. The case study in this section is based on description of the MMA design, MM behavior, comparison of measured reflection coefficient and absorptance with the simulated results of a MMA with frequency selective surface consists of concentric continuous rings (CCRs) [28]. The concentric rings are of different widths and radius and exhibits high degree of symmetry that makes the MM absorber insensitive to wide incidence angle and polarization state of incident EM waves.

5.1 Optimized design of metamaterial absorber

Figure 11 shows the optimized design of a MMA and the optimized dimensions of the unit cell and the rings are depicted in **Tables 3** and **4**, respectively. The design is optimized by parametric analysis of the dimensions of the unit cell wherein anyone parameter is varied and rest all other parameters are kept constant. This process is repeated for each and every parameter and the effect on the absorptance is observed.

5.2 Simulation of metamaterial absorber

The optimized design of the multi-band MMA is simulated to study the response of the MMA at different frequency bands depicted in **Figure 12**.

The simulated reflection coefficients and absorptance is depicted in **Table 5**. The absorptance is estimated from simulated values of reflectance $|S_{11}(f)|^2$ and transmittance $|S_{21}(f)|^2$ using the Eq. (21) [8, 9, 24, 25]. The presence of continuous metallic bottom layer blocks the transmission of incident EM waves thereby making transmittance (S_{21}) zero shown in **Figure 12**.

$$A(f) = 1 - |S_{11}(f)|^2 - |S_{21}(f)|^2 \quad (17)$$

To better understand the contribution of CCRs towards the resonance frequencies, the surface current distributions are investigated at different resonance frequencies. When an EM radiation is applied along the axis of the CCRs, depending upon the resonant properties of CCRs strong surface currents are induced either opposes or enhances the incident EM field. **Figure 13** shows the surface current induced on the surface [29] of CCRs on the top layer of MMA structure. The arrowhead indicates the

Design parameter	r_1	r_2	r_3	r_4	r_5	r_6	r_7	r_8	w	w_1	w_2	a	t
Dimension (mm)	3.93	3.68	2.35	2.10	3.20	2.85	2.35	1.93	0.25	0.35	0.42	18	1.5

Table 3.
Optimized dimensions of the unit cell.

Structure	OR-1	IR-1	OR-2	IR-2
Width of ring	0.25 mm	0.25 mm	0.35 mm	0.42 mm
Mean radius of ring	$r_{m(OR\text{-}1)}$ 3.80 mm	$r_{m(IR\text{-}1)}$ 2.22 mm	$r_{m(OR\text{-}2)}$ 3.02 mm	$r_{m(IR\text{-}2)}$ 2.14 mm

Table 4.
Optimized dimensions of the rings.

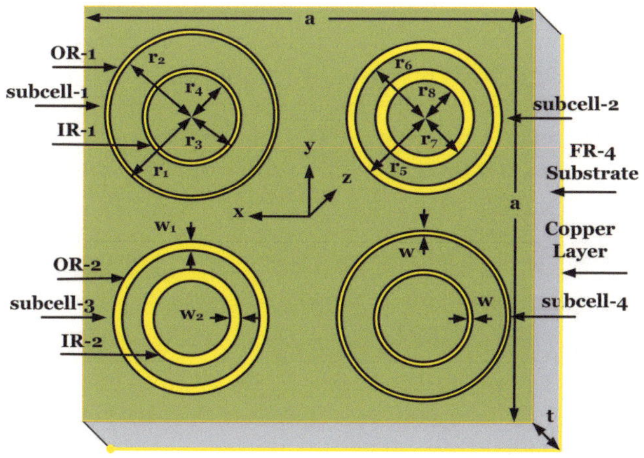

Figure 11.
The 3-D perspective view of the geometry of the unit cell with dimensions of the subcells.

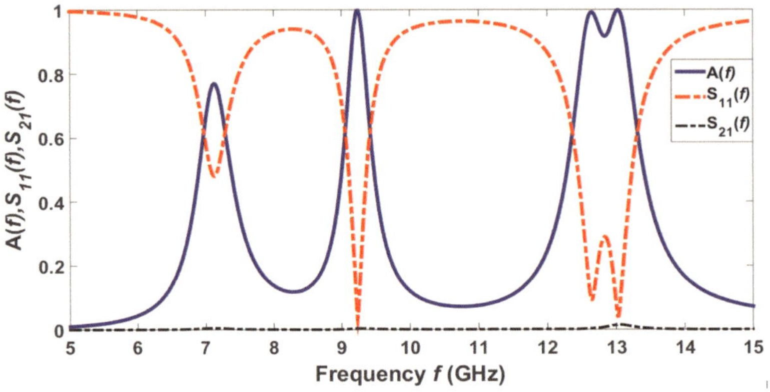

Figure 12.
Reflection coefficient $S_{11}(f)$, *Transmission coefficient* $S_{21}(f)$ *and Absorptance* A(f) *of the proposed MMA structure at 7.06, 9.18, 12.62 and 13 GHz.*

Resonance frequency	7.06 GHz	9.18 GHz	12.62 GHz	13 GHz
Reflection coefficients	−6.5 dB	−36 dB	−21.5 dB	−27.5 dB
Absorptance	78%	99.9%	99.2%	99.8%

Table 5.
Effective resonance frequencies, reflection coefficients, and absorptance.

direction of current induced and the density of arrows shows the magnitude of surface current induced. Each continuous ring acts as a resonator and contributes towards absorption of incident EM radiation by resonating at a particular frequency shown in **Figure 13**. The two-dimensional periodic array of unit cells that consisted of four pair of CCRs in MMA structure resulted in strong coupling of CCRs in the MMA structure that further contributed to high absorptance of incident EM radiation.

The circulating anti-parallel surface currents are induced from magnetic response whereas the circulating parallel surface currents are induced from electric response [29]. The surface currents resulted in coupling of incident EM radiation field with that of electric and magnetic responses and hence enhancement of localized EM field is established at resonance frequencies. At resonance frequencies the simultaneous magnetic as well as electric resonance results in complete absorptance of incident EM radiation under normalized impedance matching conditions (impedance matching of MMA with that of free space). Therefore, the incident EM radiation ranging from 5-15 GHz got confined within the MMA structure with minimum reflection and maximum absorptance.

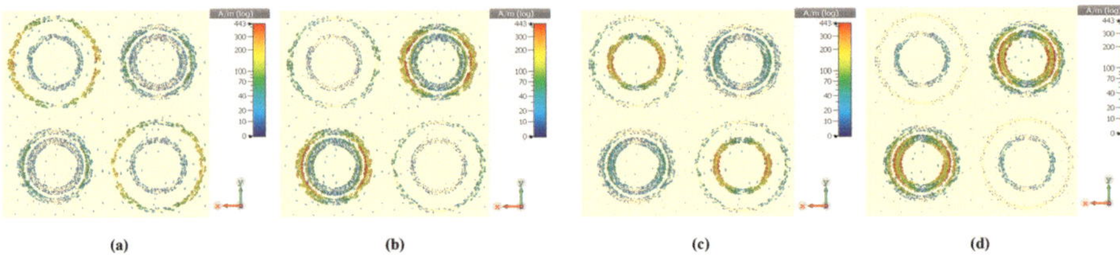

Figure 13.
Surface current distributions at resonance frequencies (a) 7.06 GHz. (b) 9.18 GHz. (c) 12.62 GHz. (d) 13 GHz.

With the incident EM radiation, the localized E-field is concentrated in x-direction along the CCRs as shown in **Figure 14**, and the localized H-field is concentrated in y-direction along the CCRs as shown in **Figure 15**. The metallic patch of CCR acts as inductive element on which the magnetic field gets concentrated. The inductive element is responsible for magnetic resonance. The electric field is concentrated in between the metallic patch of CCRs where the gaps between the metallic patches act as capacitive elements, responsible for electric resonance [3]. Thus, the magnetic and electric resonance at each CCR contributes to the absorptance peak at a particular resonance frequency. The concentration of surface current, E-field, and H-fields along a CCR indicates the responsiveness of that CCR for absorptance at a distinct resonance frequency [3]. From **Figures 13–15**, it is evident that the surface current and field distribution, is concentrated along the outer ring OR-1 and therefore it contributes to the absorptance at first resonance frequency 7.06 GHz. For second resonance frequency at 9.18 GHz the surface current and field distribution is concentrated along the outer ring OR-2 which contributes to high absorptance greater than 99%. Third and fourth resonance frequencies merged to generate the absorptance band from 12.62 to 13 GHz where the surface current and field distribution is concentrated along the inner rings IR-1 and IR-2 which contributes to absorptance greater than 99%.

5.3 Metamaterial properties of metamaterial absorber

To verify the maximum absorptance under perfectly matched conditions and dissipation of EM waves inside the MMA structure, the values of effective impedance $\tilde{Z}_{\text{eff}}(f)$, permittivity $\tilde{\varepsilon}_{\text{eff}}(f)$, permeability $\tilde{\mu}_{\text{eff}}(f)$ and refractive index $\tilde{\eta}_{\text{eff}}(f)$ are retrieved from S_{11} and S_{21} parameters using standard methods of retrieval like NRW method [25], direct refractive index (DRI) method [30].

Therefore $\tilde{Z}_{\text{eff}}(f) = \sqrt{\frac{\tilde{\mu}_{\text{eff}}(f)}{\tilde{\varepsilon}_{\text{eff}}(f)}} = 1$ at maximum absorptance frequencies. The incident EM waves propagate through the top layer of MMA with no reflections under perfectly matched conditions and blocked by the continuous metallic ground layer.

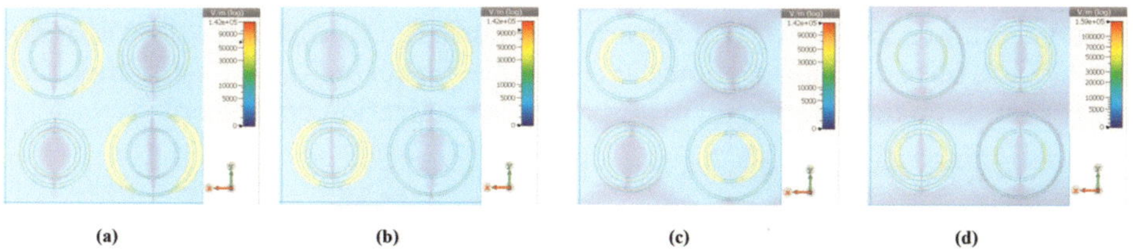

Figure 14.
E-field distributions at resonance frequencies (a) 7.06 GHz. (b) 9.18 GHz. (c) 12.62 GHz. (d) 13 GHz.

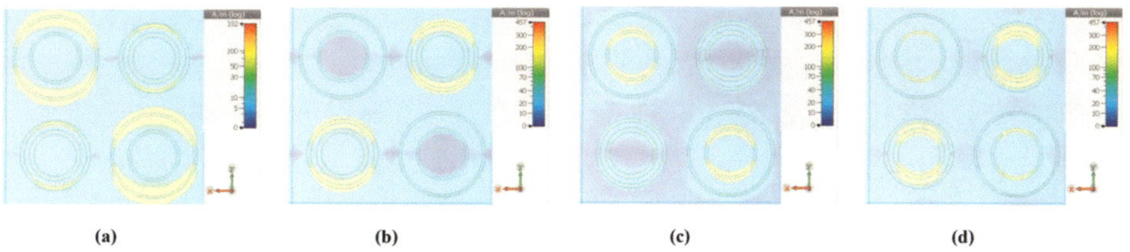

Figure 15.
H-field distributions at resonance frequencies (a) 7.06 GHz. (b) 9.18 GHz. (c) 12.62 GHz. (d) 13 GHz.

Parameter	7.06 GHz	9.18 GHz	12.62 GHz	13 GHz
$\tilde{Z}_{\text{eff}}(f)$	0.41 – j0.15	0.95 + j0.04	1.15 + j0.04	0.94 + j0.04
$\tilde{\varepsilon}_{\text{eff}}(f)$	1.59 + j6.42	0.04 + j4.01	−0.12 + j4.44	0.22 + j3.34
$\tilde{\mu}_{\text{eff}}(f)$	−0.188 + j2.45	−0.03 + j3.99	0.17 + j4.58	−0.17 + j3.32
$\tilde{\eta}_{\text{eff}}(f)$	0.2107 + j3.98	0.0288 + j4.00	0.188 + j4.51	0.0244 + j3.33

Table 6.
Values of $\tilde{Z}_{\text{eff}}$ (f), $\tilde{\varepsilon}_{\text{eff}}$(f), $\tilde{\mu}_{\text{eff}}$(f) and $\tilde{\eta}_{\text{eff}}$(f) at 7.06, 9.18, 12.62 and 13 GHz [28].

The MMA structure exhibits the large value of imaginary part of effective $\tilde{\eta}_{\text{eff}}(f)$ with respect to value of real part of effective $\tilde{\eta}_{\text{eff}}(f)$ $[\eta''_{\text{eff}}(f) \gg \eta'_{\text{eff}}(f)]$, which results in exponential decay of EM waves inside the MMA (**Table 6**).

5.4 Measured reflection coefficient and absorptance of metamaterial absorber

A 10 × 10 two-dimensional periodic array consisting of 100 unit cells was fabricated on FR-4 substrate (ε_r = 4.4, tan δ = 0.02) of thickness 1.5 mm by UV photolithography method and wet etching process is shown in **Figure 16a**. The experimental process is calibrated, and the measurements recorded for the calibration process are shown **Figure 16b**. The calibrated reflected power is the difference of reflected power measured from the bottom copper laminated side and the reflected power measured from the top layer comprising of CCRs by the VNA through receiving antenna in terms of (S_{11}) parameter as described in Section 3.3.

The measured resonance frequencies, the reflection coefficients S_{11} and the absorptance peaks at these resonance frequencies are in close approximation with the simulated values. The reflection coefficients in terms of S_{11} parameter are measured experimentally and it is found that at first resonance frequency (7.20 GHz) the S_{11} value is −12.7 dB. At second resonance frequency (9.3 GHz) the S_{11} value is measured to be −17.4 dB. At third resonance frequency (12.61 GHz) the S_{11} value is measured to be −13.1 dB. At fourth resonance frequency (13.07 GHz) the S_{11} value is measured to be −21.8 dB shown in **Figure 17a**. **Figure 17b** shows the comparison of estimated simulated and measured absorptance at resonance frequencies. The estimated absorptance at first resonance frequency from measured reflection coefficients is 94.5%.

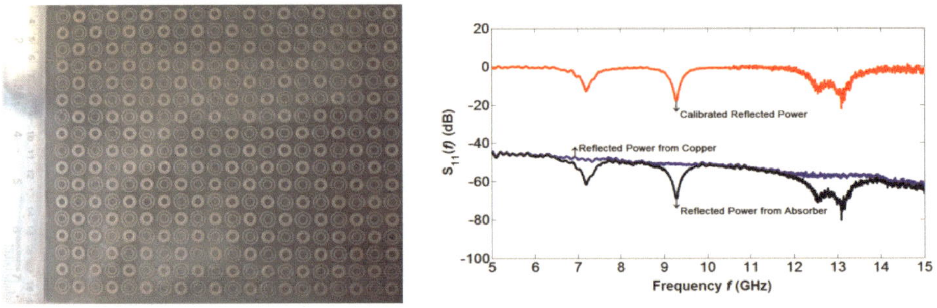

(a) (b)

Figure 16.
(a) Fabricated MMA structure with two-dimensional periodic array comprising of 100 unit cells within 180 mm by 180 mm. (b) Measured reflected power as a difference of reflected power from bottom copper laminated layer and top layer with CCRs of MMA structure from 5 to 15 GHz.

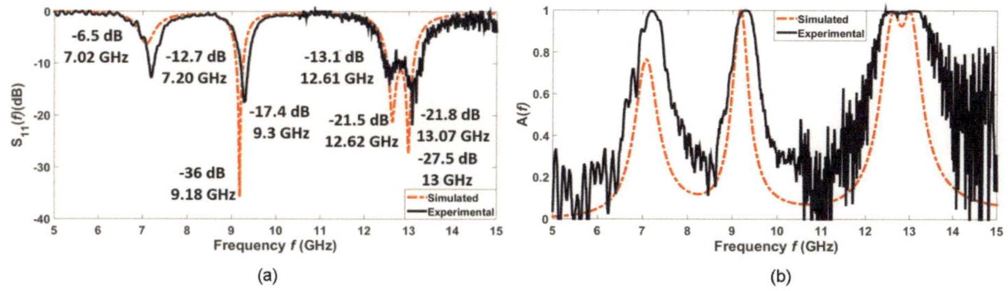

Figure 17.
(a) Comparison between simulated and experimentally measured values of reflection coefficient $S_{11}(f)$*. (b) Estimated simulated and measured absorptance at resonance frequencies for MMA at 7.20, 9.3, 12.61 and 13.07 GHz.*

At second resonance frequency the measured absorptance is 97.9%. At third resonance frequency the measured absorptance is 95.1%. At fourth resonance frequency the measured absorptance is 99.3%. The third and fourth resonance peaks merged together to generate a 3 dB broad-band of bandwidth 1.2 GHz that lies in *K*u-band. The measured resonance frequencies are found to be slightly shifted towards right (at higher frequency) as compared to the simulated resonance frequencies. This slight variation is contributed by fractional variation in the width of the CCRs structure fabricated on an area of 180 × 180 mm by UV photolithography and wet etched process as compared to dry etched process of fabrication.

6. Conclusions

MMs are artificially engineered materials that exhibit negative permittivity and permeability. These materials possess unusual EM properties that make these materials unique. The interest in MMs increased with first demonstration of MM absorber by Landy et al. in 2008. The conventional EM absorbers were limited in performance due to their bulky size and the performance is limited to microwave frequencies due to impedance matching conditions. One of the application domains of MMs is a MPA. The frequency selective top layer of conventional MPA has matched impedance with that of the free space. Under perfectly matched conditions the incident EM waves are absorbed with minimum reflections and get attenuated inside the substrate. The MM design, MM behavior, impedance matched conditions, and absorptance of EM waves under perfectly matched conditions are explained with a case study of a multi-band MPA that finds application in stealth technology.

Author details

Alkesh Agrawal
Shri Ramswaroop Memorial University, Barabanki, India

*Address all correspondence to: alkesh.agrawal26@gmail.com

References

[1] Veselago VG. The electrodynamics of substances with simultaneously negative values of ε and μ. Soviet Physics Uspekhi. 1968;**10**(4):509-514. DOI: 10.1070/PU1968v010n04ABEH003699

[2] Shelby RA, Smith DR, Schultz S. Experimental verification of a negative index of refraction. Science. 2001;**292**(5514):77-79. DOI: 10.1126/science.1058847

[3] Smith DR, Padilla WJ, Vier DC, Nemat-Nasser SC, Schultz S. Composite medium with simultaneously negative permeability and permittivity. Physics Review Letters. 1999;**84**(10):4184-4187. DOI: 10.1103/PhysRevLett.84.4184

[4] Pendry JB, Holden AJ, Stewart WJ, Youngs I. Extremely low frequency plasmons in metallic microstructures. Physics Review Letters. 1996;**76**(25): 4773-4776. DOI: 10.1103/PhysRev Lett.76.4773

[5] Walser RM. Electromagnetic metamaterials. In: Proceeding SPIE Complex Mediums II beyond Linear Isotropic Dielectrics, San Diego. 2001; 4467: 1–15. DOI: 10.1117/12.432921

[6] Pendry JB. Negative refraction makes a perfect lens. Physics Review Letters. 2000;**85**(18):3966-3969. DOI: 10.1103/PhysRevLett.85.3966

[7] Fang N, Lee H, Sun C, Zhang X. Sub-diffraction-limited optical imaging with a silver super lens. Science. 2005; **308**:534-537. DOI: 10.1126/science.1108759

[8] Landy NI, Sajuyigbe S, Mock JJ, Smith DR, Padilla WJ. Perfect metamaterial absorber. Physical Review Letters. 2008;**100**:207402. DOI: 10.1103/PhysRevLett.100.207402

[9] Tao H, Landy NI, Bingham CM, Zhang X, Averitt RD, Padilla WJ. A metamaterial absorber for the terahertz regime: design, fabrication and characterization. Optics Express. 2008;**16**: 7181-7188. DOI: 10.1364/OE.16.007181

[10] Pendry JB, Schurig D, Smith DR. Controlling electromagnetic fields. Science. 2006;**312**:1780-1782. DOI: 10.1126/science.1125907

[11] Schurig D, Mock JJ, Justice BJ, Cummer SA, Pendry JB, Starr AF, et al. Metamaterial electromagnetic cloak at microwave frequencies. Science. 2006; **314**:977-980. DOI: 10.1126/science.1133628

[12] Rahm M, Schurig D, Roberts DA, Cummer SA, Smith DR, Pendry JB. Design of electromagnetic cloaks and concentrators using form-invariant coordinate transformations of Maxwells equations. Photonics and Nanostructures. 2008;**6**(1):87-95. DOI: 10.1016/j.photonics.2007.07.013

[13] Kim T, Seo C. A novel photonic bandgap structure for low-pass filter of wide stop-band. IEEE Microwave Guided Wave Letters. 2000;**10**:13-15. DOI: 10.1109/75.842072

[14] Baena JD, Marqués R, Medina F. Artificial magnetic metamaterial design using spiral resonators. Physical Review B. 2004;**69**:014402

[15] Ziolkowski RW. Metamaterial-based efficient magnetically small antennas. IEEE Transactions on Antennas and Propagation. 2006;**54**(7):2113-2130. DOI: 10.1109/TAP.2006.877179

[16] Upadhyay KK, Agrawal A, Misra M. Wide-band log-periodic microstrip

antenna with defected ground structure for C-band applications. Progress in Electromagnetics Research C. 2021; **112**:127-137. DOI: 10.2528/PIERC21 031106

[17] Agrawal A, Misra M. Angle of incidence and polarization insensitive ultra- thin ultra wide-band metamaterial absorber based on novel non-linearity in unit cell design. Journal of Microwaves, Optoelectronics and Electromagnetic Applications. 2021;**20**(2):556-569. DOI: 10.1590/2179-10742021v20i31207

[18] Salisbury WW. Absorbent body for electromagnetic waves. United States Patent 2599944. 1954

[19] Berenger JP. A perfectly matched layer for the absorption of electromagnetic waves. Journal of Computational Physics. 1994;**114**(2): 185-200. DOI: 10.1006/jcph.1994.1159

[20] Watts CM, Liu XL, Padilla WJ. Metamaterial electromagnetic wave absorbers. Advanced Materials. 2012;**24**: 98-120. DOI: 10.1002/adma.201200674

[21] Costa F, Monorchio A, Manara G. Analysis and design of ultra thin electromagnetic absorbers comprising resistively loaded high impedance surfaces. IEEE Transactions on Antenna and Propagation. 2010;**58**(5):1551-1558. DOI: 10.1109/TAP.2010.2044329

[22] A. Motevasselian and Jonsson BLG. Radar cross section reduction of aircraft wing front end. In: Proceedings IEEE International Conference on Electromagnetics in Advanced Applications (ICEAA '09), pp. 237–240, Turin, Italy, 2009

[23] Liu X, Tyler T, Starr T, Starr AF, Jokerst NM, Padilla WJ. Taming the blackbody with infrared metamaterials as selective thermal emitters. Physical Review Letters. 2011;**107**:045901. DOI: 10.1103/PhysRevLett.107.045901

[24] Liu X, Starr T, Starr AF, Padilla WJ. Infrared spatial and frequency selective metamaterial with near-unity absorbance. Physical Review Letters. 2010;**104**:207403. DOI: 10.1103/PhysRev Lett.104.207403

[25] Luukkonen O, Maslovski SI, Tretyakov SA. A stepwise Nicolson-Ross-Weir based material parameter extraction method. IEEE Antenna Wireless Propogation Letters. 2011;**10**: 1295-1298. DOI: 10.1109/LAWP.2011. 2175897

[26] Hao J, Wang J, Liu X, Padilla WJ, Zhou L, Qiu M. High performance optical absorber based on a plasmonic metamaterial. Applied Physics Letters. 2010;**96**(25):251104

[27] Liu N, Mesch M, Weiss T, Hentschel M, Giessen H. Infrared perfect absorber and it's application as plasmonic sensor. Nano Letters; **10,** 7:2342-2348. DOI: 10.1021/ nl9041033

[28] Agrawal A, Misra M, Singh A. Oblique incidence and polarization insensitive multiband metamaterial absorber with quad paired concentric continuous ring resonators. Progress in Electromagnetics Research M. 2017;**60**: 33-46. DOI: 10.2528/PIERM17061302

[29] Campbell SD, Ziolkowski RW. Lightweight, flexible, polarization-insensitive, highly absorbing meta-films. IEEE Transactions on Antennas and Propagation. 2013;**61**(3):1191-1200. DOI: 10.1109/TAP.2012.2227658

[30] Islam SS, Faruque MRI, Islam MT. A new direct retrieval method of refractive index for the metamaterial. Current Science. 2015;**109**:337-342

Chapter 6

Amplitude-Dependent Acoustic Absorber

Jiangyi Zhang

Abstract

In this chapter, we consider the design of 1D amplitude-dependent acoustic absorber, i.e., acoustic metamaterial composed of an air-filled waveguide periodically side-loaded by holes. Firstly we present experimental results about high-amplitude sound wave propagation in the structure. We find that there is not only the linear viscothermal and radiation losses, but also the nonlinear losses. High-amplitude sound waves at the locations of the side holes could introduce nonlinear losses, which is due to the geometrical discontinuity. This phenomenon could be found in experiments about amplitude-dependent reflection, transmission, and absorption coefficients. The experimental results show the possibility to use the system as a nonlinear absorber, that is, nonlinear losses change the nature of the device from a reflective to an absorbing one. Our results reveal the role of nonlinear losses in the proposed device and also provide a quite accurate analytical model to capture the effect of such losses. In the end, we analytically tune the parameters of the device and design 1D amplitude-dependent acoustic absorber.

Keywords: nonlinear losses, nonlinear acoustic absorber, acoustic metamaterial, transfer matrix method, acoustic waveguide

1. Introduction

In the nature, the properties of conventional materials depend on the composing atoms/molecules and the chemical bonds, while their physical properties such as mass density and bulk modulus are always positive. Metamaterials, namely periodic or random man-made structures composed of meta-atoms, for example, sub-wavelength resonators the size of which is bigger than the atomic scale but much smaller than the relevant radiated wavelength, are designed to exhibit exotic properties not commonly found in nature. The physical properties of metamaterials are described by effective parameters which could be positive, negative, near-zero or approaching-infinity [1]. The concept of metamaterials was firstly theorized by the Russian physicist Veselago [2] for electromagnetic waves in 1968. He considered the possible existence of metamaterials with simultaneously negative permittivity and permeability, i.e., left-handed materials or double negative metamaterials. In 2000, Liu et al. firstly introduced the concept of acoustic metamaterials [3]. Later, acoustic metamaterials have been widely developed to deeply control and manipulate waves due to their exotic properties not found in nature. As a result, a plethora of applications has been developed, including acoustic diodes [4, 5], acoustic lenses for sub-diffraction imaging [6], acoustic sound focusing based on gradient index lenses [7–9], acoustic topological systems [10–12] and acoustic cloaking [13–15].

Wave propagation in acoustic metamaterials is subject to losses such as visothermal losses at the solid-fluid interface [16], which can have deleterious consequences. For example, Heríquez et al. [17] found that viscothermal losses could destroy the predicted behavior of double negative acoustic metamaterials designed by Gracía-Salgado et al. [18], and they even speculated that rigid-based metamaterials could become absorbers in particular situations. Previously, Guild et al. [19] and Romero-Garcá et al. [20] proposed specific acoustic metamaterials absorbers based on dissipative effect. Losses could also change the dispersion relation of the system. For example, Theocharis et al. [21] show that near-zero group velocity dispersion band disappears in the presence of the viscothermal losses. Thus, losses are non-negligible in such kind of materials and should be not ignored. Perfect and broadband absorber is an important branch of acoustic metamaterials [20, 22].

In the field of metamaterials, only few studies exist regarding nonlinear effects. In 2003, Lapine et al. proposed the concept of nonlinear metamaterials in the field of electromagnetism [23]. During the last years, the study of nonlinear acoustic metamaterials received increased attention [24–26]. In the case of acoustic metamaterials, the presence of nonlinearity is easily introduced at high acoustic levels. Accompanying with wave propagation, the localized pressure changes. High pressure increases the local temperature, and the local speed of sound in the compressible material is proportional to the temperature [27]. As a result, the wave with high pressure phase travels faster compared to a wave of low pressure phase. During the nonlinear wave propagation, higher harmonics are generated. Dispersion effect introduced by the resonators and periodicity of the acoustic metamaterials could tailor the source and generated harmonics. Thus, one can observe the beating of higher generated harmonics [28–30]. Acoustic diodes have been also designed based on the combination of nonlinearity and dispersion. Liang et al. realized the nonreciprocal acoustic transmission by breaking the linearity, which violates some assumptions of Onsager-Casimir principle [4]. In granular crystals, Boechler et al. envisaged the bifurcation-based acoustic switching and rectification [31]. Devaux et al. achieved asymmetric acoustic propagation by a nonlinear self-demodulation effect [32]. There are many other nonlinear processes, for example, the nonlinear dispersion relation [33], nonlinear acoustic lenses [34], waves coupling in nonlinear metamaterials [35], discrete breathers [36], the self-demodulation effect [37, 38], and the formation of acoustic solitons of various types-pulse-like [39, 40] and envelope ones [40, 41], namely robust localized waves propagating undistorted due to a balance between dispersion and nonlinearity.

Despite this extensive body of nonlinear acoustic metamaterials and acoustic absorbers, only few works consider nonlinear losses. However, in acoustics, nonlinear losses may easily appear, e.g., due to the geometrical discontinuity [42, 43]. In this article, we design 1D amplitude-dependent acoustic absorber made by an air-filled waveguide periodically side-loaded by holes with sharp edges. These sharp edges can be considered as geometrical discontinuities, which could introduce nonlinear losses. Based on the accurate analytical model which could predict the experimental results, we tune the parameters of the device in order to design a nonlinear perfect absorber. Particularly the industry is concerned about the perfect absorber in low frequencies. However, nonlinear losses could change the nature of the device from reflective one to an absorber, especially in low frequencies.

This chapter is structured as follows. In Section 2, we introduce the setup and the method of processing the experimental data. In Section 3, we derive the analytical model based on transfer matrix method. In Section 4, we discuss the experimental phenomenon and analytical model. Based on results obtained, we tune the parameters of the device in order to design an 1D amplitude-dependent acoustic perfect absorber in low frequencies. Finally Section 5 summarizes the findings and discusses future research directions.

2. Experiments

2.1 Experimental set-up

Figure 1 shows the experimental set-up in order to characterize the amplitude-dependent reflection, transmission and absorption coefficient. This system is an air-filled acoustic cylindrical waveguide periodically loaded by side-holes. The length of the waveguide is 3 m; the internal radius is $r = 0.025$ m; there is 40 side-holes; the distance between two side holes is $d = 0.05$ m; the radius of the side holes is $r_H = 0.0035$ m and the length is $l_H = 0.005$ m. We use white noise with frequencies from 200 Hz to 1000 Hz as source. In order to avoid back reflections in the waveguide, we use a properly designed anechoic termination [21, 45], and do the experiments in an anechoic room.

We perform four experiments with different source amplitudes. Here we use a four microphone method [46] to obtain the transmission, reflection and absorption coefficient. We put two microphones (1/2 inch *B&K*) at the one end of the wave-guide; and another two microphones on other side of the structure, shown in **Figure 1**. This method allows the measurement of both the forward and backward waves for the fundamental frequency.

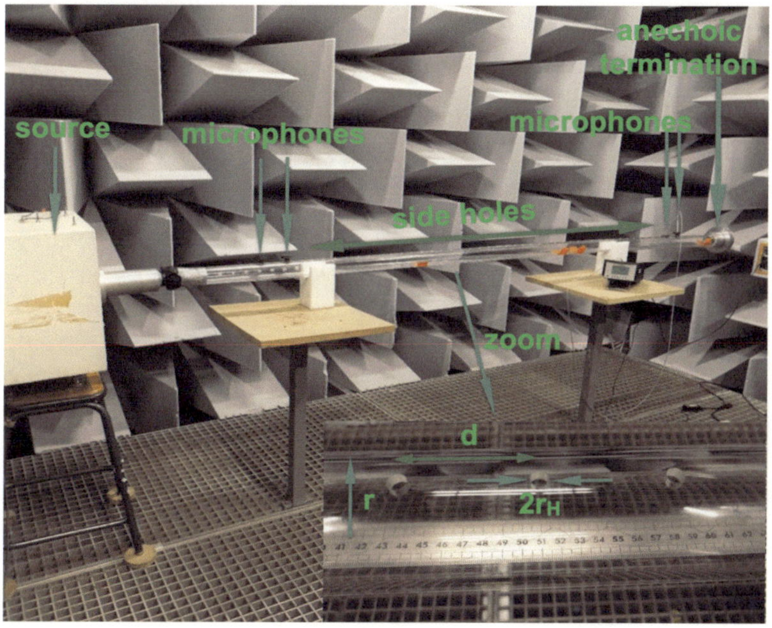

Figure 1.
(color online) picture of the experimental setting for the measurement of the amplitude-dependent reflection, transmission and absorption coefficients, adapted from ref. [44].

2.2 Experimental data processing

Here are the details of four microphone method. p_1, p_2, p_3 and p_4 present the complex sound pressure measured by the microphones at x_1, x_2, x_3 and x_4 in **Figure 2**, with $x_1 = 0.2$ m, $x_2 = 0.1$ m, $x_3 = 2.1$ m and $x_4 = 2.2$ m respectively. Acoustic pressures measured by the microphone comprise various superpositions of positive- and negative-going plane waves in the waveguide,

$$
\begin{aligned}
p_1 &= \left(Ae^{-jkx_1} + Be^{jkx_1}\right)e^{i\omega t},\\
p_2 &= \left(Ae^{-jkx_2} + Be^{jkx_2}\right)e^{i\omega t},\\
p_3 &= \left(Ce^{-jkx_3} + De^{jkx_3}\right)e^{i\omega t},\\
p_4 &= \left(Ce^{-jkx_4} + De^{jkx_4}\right)e^{i\omega t},
\end{aligned}
\tag{1}
$$

where A and B (C and D) are the amplitudes of the positive- and negative going plane wave at the beginning (ending) of the waveguide respectively, which could be derived from Eq. (1),

$$
\begin{aligned}
A &= \frac{j\left(p_1e^{jkx_2} - p_2e^{jkx_1}\right)}{2\sin k(x_1 - x_2)},\\
B &= \frac{j\left(p_2e^{-jkx_1} - p_1e^{-jkx_2}\right)}{2\sin k(x_1 - x_2)},\\
C &= \frac{j\left(p_3e^{jkx_4} - p_4e^{jkx_3}\right)}{2\sin k(x_3 - x_4)},\\
D &= \frac{j\left(p_4e^{-jkx_3} - p_3e^{-jkx_4}\right)}{2\sin k(x_3 - x_4)},
\end{aligned}
\tag{2}
$$

and provide the input data for subsequent transfer matrix calculations.

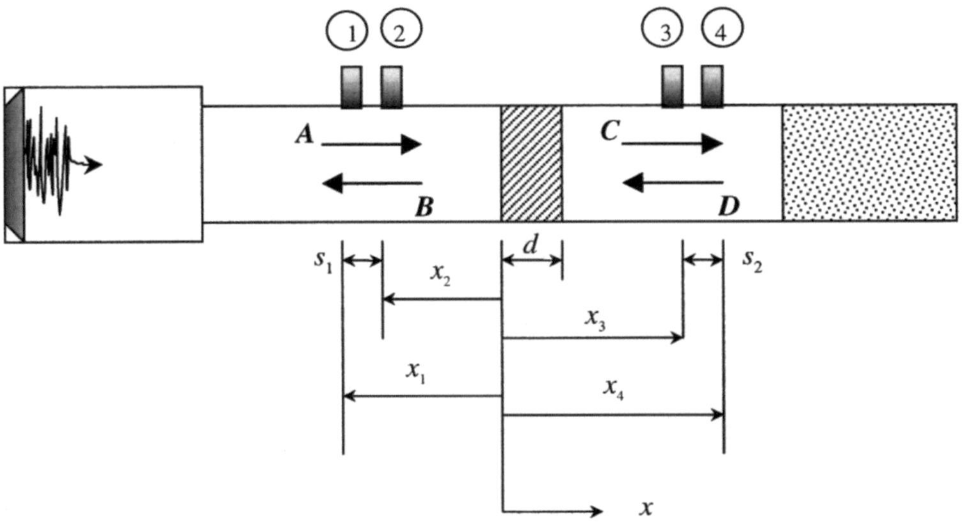

Figure 2.
Schematic diagram of the standing wave tube, adapted from ref. [46].

To relate the sound pressures and velocities on the two faces of 40 side holes, transfer matrix is used, extending from $x = 0$ to $x = 40 * d$, as shown in **Figure 2**, i.e.,

$$\begin{bmatrix} p \\ v \end{bmatrix}_{x=0} = \begin{bmatrix} T_{11} & T_{12} \\ T_{21} & T_{22} \end{bmatrix} \begin{bmatrix} p \\ v \end{bmatrix}_{x=40d}, \tag{3}$$

where $p|_{x=0}$, $v|_{x=0}$, $p|_{x=40d}$ and $v|_{x=40d}$ may easily be expressed in terms of the positive- and negative-going plane wave component amplitude, i.e.,

$$p\Big|_{x=0} = A + B, v\Big|_{x=0} = \frac{A - B}{\rho_0 c} p\Big|_{x=40d} = Ce^{-jk40d} + De^{jk40d}, v\Big|_{x=40d} = \frac{Ce^{-jk40d} - De^{jk40d}}{\rho_0 c}. \tag{4}$$

Eq. (3) presents two equations with four unknowns, T_{11}, T_{12}, T_{21} and T_{22}. In order to solve the transfer matrix elements, we need two additional equations, which could be generated by reciprocity and symmetry nature of the material. Pierce noted that reciprocity requires that the determinant of the transfer matrix be unity [46, 47],

$$T_{11}T_{22} - T_{12}T_{21} = 1,$$

and for symmetrical system,

$$T_{11} = T_{22}.$$

Then the transfer matrix elements are solved,

$$\begin{aligned} T_{11} &= \frac{p|_{x=40d} v|_{x=40d} + p|_{x=0} v|_{x=0}}{p|_{x=0} v|_{x=40d} + p|_{x=40d} v|_{x=0}}, \\ T_{12} &= \frac{p|_{x=0}^2 - p|_{x=40d}^2}{p|_{x=0} v|_{x=40d} + p|_{x=40d} v|_{x=0}}, \\ T_{21} &= \frac{v|_{x=0}^2 - v|_{x=40d}^2}{p|_{x=0} v|_{x=40d} + p|_{x=40d} v|_{x=0}}, \\ T_{22} &= \frac{p|_{x=40d} v|_{x=40d} + p|_{x=0} v|_{x=0}}{p|_{x=0} v|_{x=40d} + p|_{x=40d} v|_{x=0}}, \end{aligned} \tag{5}$$

which provide the input data for subsequent reflection and transmission coefficients calculations.

In order to calculate reflection and transmission coefficients, we assume that the incidental plane wave has unit amplitude, and the termination is anechoic, so that we could consider the parameter D is negligible compared to parameter C. Thus we could write,

$$p\Big|_{x=0} = 1 + R, v\Big|_{x=0} = \frac{1 - R}{\rho_0 c} p\Big|_{x=40d} = Te^{-jk40d}, v\Big|_{x=40d} = \frac{Te^{-jk40d}}{\rho_0 c}, \tag{6}$$

where $R = B/A$ and $T = C/A$ are the reflection and transmission coefficients respectively. When we substituted Eq. (6) into Eq. (3), the transmission T and reflection coefficient R for the case of an anechoic termination, can be respectively expressed as,

$$T = \frac{2e^{jkd}}{T_{11} + (T_{12}/\rho_0 c) + \rho_0 c T_{21} + T_{22}}, \tag{7}$$

$$R = \frac{T_{11} + (T_{12}/\rho_0 c) - \rho_0 c T_{21} - T_{22}}{T_{11} + (T_{12}/\rho_0 c) + \rho_0 c T_{21} + T_{22}}, \tag{8}$$

and the absorption coefficient is

$$\alpha = 1 - T^2 - R^2. \tag{9}$$

Through the above data processing, we obtain the experimental amplitude-dependent reflection, transmission and absorption coefficients. The corresponding results are shown in Section 4.

3. Transfer matrix method

In this section, we analytically use transfer matrix method to capture the amplitude-dependent reflection, transmission and absorption coefficients. The transfer matrix T used to relate the sound pressures and velocities on the two faces of the acoustic waveguide with the N side holes, can be written as

$$T = \prod_{n=1}^{N} T_n \tag{10}$$

where

$$\begin{aligned} T_n = &\begin{bmatrix} \cos\left(\frac{k_{ph}d}{2}\right) & iZ_c \sin\left(\frac{k_{ph}d}{2}\right) \\ \frac{i}{Z_c}\sin\left(\frac{k_{ph}d}{2}\right) & \cos\left(\frac{k_{ph}d}{2}\right) \end{bmatrix} \begin{bmatrix} 1 & 0 \\ \frac{1}{Z_{H(n)}} & 1 \end{bmatrix} \\ &\times \begin{bmatrix} \cos\left(\frac{k_{ph}d}{2}\right) & iZ_c \sin\left(\frac{k_{ph}d}{2}\right) \\ \frac{i}{Z_c}\sin\left(\frac{k_{ph}d}{2}\right) & \cos\left(\frac{k_{ph}d}{2}\right) \end{bmatrix}, \end{aligned} \tag{11}$$

with k_{ph} and Z_c being the wavenumber and acoustic characteristic impedance in the waveguide respectively, which are given by:

$$k_{ph} = \frac{\omega_{ph}}{c_0}\left[1 + \frac{1-i}{\sqrt{\omega_{ph}\rho_0 r^2/\eta}}\left(1 + \frac{\gamma - 1}{\sqrt{Pr}}\right)\right], \tag{12}$$

$$Z_c = \frac{\rho_0 c_0}{\pi r^2}\left[1 + \frac{1-i}{\sqrt{\omega_{ph}\rho_0 r^2/\eta}}\left(1 + \frac{\gamma - 1}{\sqrt{Pr}}\right)\right]. \tag{13}$$

Here, γ is the specific heat ratio, Pr is the Prandtl number, η is the shear viscosity, while $Z_{H(n)}$ is the impedance of n-th side hole. Based on the experimental results, we conclude that the nonlinear losses, appearing at high level sound pressures due to jet and vortices formation at the locations of side holes, must be taken into account, i.e., here, $Z_{H(n)}$ is the sum of a linear impedance, Z_L, and a nonlinear one, $Z_{NL(n)}$, namely:

$$Z_{H(n)} = Z_L + Z_{NL(n)}. \tag{14}$$

On the one hand, the linear impedance of the hole can be expressed as a combination of the impedance of the hole plus the radiation impedance [48],

$$Z_L = iZ_{cH} \tan\left[k_H^{ph}(l_H + \Delta l_{Hi} + \Delta l_{Ho})\right] - \frac{1}{2}Z_{cH}\left(k_H^{ph} r_H\right)^2, \tag{15}$$

where Δl_{Hi} and Δl_{Ho} are length corrections due to the radiation inside the side holes and the radiation to the outer environment, while k_H^{ph} and Z_{cH} are the wavenumber and acoustic characteristic impedance in the side holes, obtained from Eq. (12) and Eq. (13) by using r_H instead of r. On the other hand, the nonlinear part of the impedance of n-th hole, $Z_{NL(n)}$, can be written as [42, 43],

$$Z_{NL(n)} = \beta_H Z_{cH} M_{(n)} St_{(n)}^{1/3}. \tag{16}$$

where β_H is a fitting parameter, which can be directly obtained by experiments. In addition, $M_{(n)}$ is the acoustic Mach number for n-th side hole, given by:

$$M_{(n)} = \frac{v_n}{c_0} = \frac{u_n/S_H}{c_0}, \tag{17}$$

with $u_n = v_n S_H$ being the volume velocity in the n-th side hole, while $St_{(n)}$ is the acoustic Strouhal number [42, 43, 49].

$$St_{(n)} = \frac{\omega_{ph} r_H}{v_n} = \frac{\omega_{ph} r_H}{u_n/S_H}. \tag{18}$$

Therefore, through $M_{(n)}$ and $St_{(n)}$, the nonlinear part $Z_{NL(n)}$ of the n-th side hole impedance depends explicitly on the local volume velocity u_n in n-th hole. However, the volume velocity is related to the impedance through

$$u_n = \frac{p_{(n)}}{Z_L + Z_{NL(n)}}. \tag{19}$$

Thus, we have to make some approximations, and apply an iterative method to calculate u_n. At the N-th hole, i.e., the last hole away from the source, the local volume velocity is the smallest compared to others, hence we can assume at first that the nonlinear part is zero at this position, $Z_{NL(n=N)} = 0$. From this hypothesis, we obtain the local volume velocity, $u_{(n=N)} = p_{(n=N)}/Z_L$ at this location. Here, it is noted that $p_{(n=N)}$ can be calculated by the following matrix,

$$\begin{bmatrix} p_{(n=N)} \\ u_{\mathrm{w}(n=N)} \end{bmatrix} = \begin{bmatrix} \cos\left(k_{ph}D\right) & iZ_c \sin\left(k_{ph}D\right) \\ \frac{i}{Z_c}\sin\left(k_{ph}D\right) & \cos\left(k_{ph}D\right) \end{bmatrix} \begin{bmatrix} p^+ \\ p^+/Z_c \end{bmatrix}, \tag{20}$$

where p^+ is positive-going wave at the end of waveguide (there is no negative-going wave due to the presence of the anechoic termination). Therefore,

$$p^+ = \frac{i\left(p_{M-1}e^{-ik_{ph}x_M} - p_M e^{-ik_{ph}x_{M-1}}\right)}{2\sin\left[k_{ph}(x_{M-1} - x_M)\right]} e^{-ik_{ph}\frac{x_{M-1}+x_M}{2}}, \tag{21}$$

where p_{M-1} and p_M are acoustic pressures measured at x_{M-1}, x_M by the microphone near the anechoic termination, p^+ is located at the center of p_{M-1} and p_M [i.e., at $(x_{M-1} + x_M)/2$], D is the distance between $p_{(n=N)}$ and p^+, and $u_{w(n=N)}$ is volume velocity in the waveguide.

This way, we obtain the local volume velocity, $u_{(n=N)}$, and substitute it into Eqs. (16)-(18). Then, the nonlinear impedance of the last hole, as well as the transfer matrix $T_{n=N}$ of last cell, can easily be derived. A similar iterative calculation can be applied considering that, in each iteration, first we approximate $Z_{NL(n)} = Z_{NL(n+1)}$. This procedure allows us to calculate the local volume velocity in the hole, and then the nonlinear impedance of the hole.

Up to now, we have the space-dependent nonlinear impedance of holes with different amplitudes of wave excitations. Then we can derive the transfer matrix of each cell, T_n, as well as the total transfer matrix, T, see Eqs. (10) and (11). Analytical reflection, transmission and absorption coefficients can be found by the total transfer matrix T.

4. Results and discussion

In this section, we start by discussing the experimental results. Firstly, we perform the experiment with a small-amplitude source. Here we choose 100 dB (linear limit), see black circles in **Figure 3**. Then, we change the amplitude of the source, for example, 120 dB (green squares), 130 dB (blue stars) and 140 dB (red triangles). The bigger the amplitude of the source is, the stronger the nonlinear effect, as well as nonlinear losses are. As a results, when the amplitude of the source is increased, maximum value of the absorption coefficient increases, and its bandwidth broadens. Here we have to highlight that in the low frequency, for example 200 Hz, the absorption coefficient in the case of 140 dB is 8 times larger than that in the case of 100 dB. These preliminary results pave the way to design the nonlinear acoustic absorber.

These experimental phenomenon could be analytically explained by the transfer matrix method, as shown in Section 3, using a nonlinear impedance model for the side-holes (see solid lines in **Figure 3**). Here we have to mention that β_H is a fitting parameter Eq. (16), because the nonlinear impedance of hole also depends on the amplitude of the wave excitation: the larger the amplitude, the larger the nonlinear impedance of holes, i.e., the larger β_H. In the linear limit (small source amplitude), there are no nonlinear losses for holes, i.e., $\beta_H = 0$. We use these rules to find the value of this fitting parameter, β_H, by comparing directly with the experimental results. In Refs. [50, 51], β_H is determined by means of numerical simulations: Disselhorst and Wijngaarden [50] found values of β_H between 0.6 and 1.0, while Peters and Hirschberg [51] evaluate β_H to 0.2. In our case, by direct comparison with

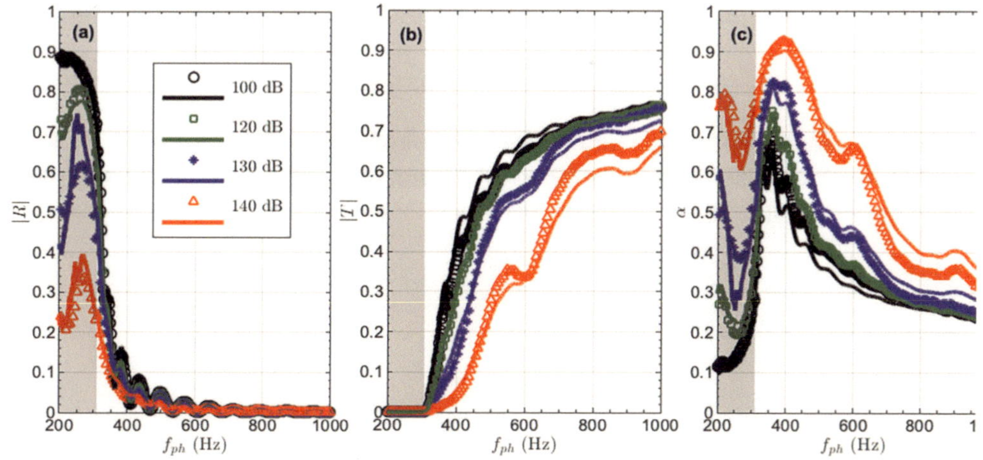

Figure 3.
(color online) amplitude of the (a) reflection coefficients, (b) transmission coefficients and (c) absorption coefficients adapted from ref. [44]. Black circles (lines), green squares (lines), blue stars (lines) and red triangles (lines) present the experimental (analytical) results with a source level around 100 dB, 120 dB, 130 dB, and 140 dB, respectively.

experiments, we found that β_H depends on the amplitude as $\beta_{H(140\text{ dB})} = 0.6$, $\beta_{H(130\text{ dB})} = 0.4$, $\beta_{H(120\text{ dB})} = 0.3$, while $\beta_{H(100\text{ dB})} = 0$ (linear case), i.e., the side holes feature no nonlinear losses when the amplitude of the source is small ($\lesssim$100 dB). In the experiments, we have also observed the creation of flow through the holes, which led us to correct the length l_H by adding 0.0005 m. This value is independent of the amplitude. Comparing to other works [52–54], we have: $l_H = 0.005$ m, $\Delta l_{Hi} = 0.0027$ m and $\Delta l_{Ho} = 0.0026$ m, i.e., $l_H + \Delta l_{Hi} + \Delta l_{Ho} = 0.0103$ m; hence, the value of 0.0005 m is small, but cannot be ignored (**Figures 3** and **4**).

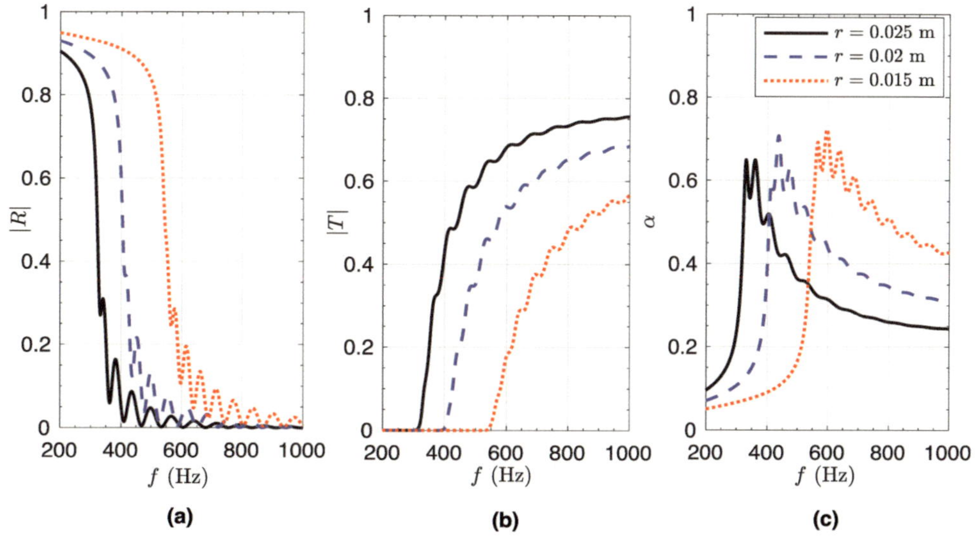

Figure 4.
(color online) amplitude of the (a) reflection coefficients, (b) transmission coefficients and (c) absorption coefficients. Black lines ($r = 0.025$ m), blue dashed lines ($r = 0.02$ m), red dotted lines ($r = 0.015$ m) present the analytical results with a source level around 100 dB (linear case).

Now we tune the parameters of the device in order to design an 1D amplitude-dependent acoustic perfect absorber. Here, we choose a source level around 140 dB. Firstly, when we decrease the radius of the waveguide, see **Figure 5**, the amplitude and the width of the absorption coefficient increase. Especially in the range of 400 – 600 Hz, the absorption coefficient is close to 1. We also found that decreasing the length of the side holes (see **Figure 6**) or increasing the radius of the side holes (see **Figure 7**) could improve the sound absorption of the meta-material. Finally, by

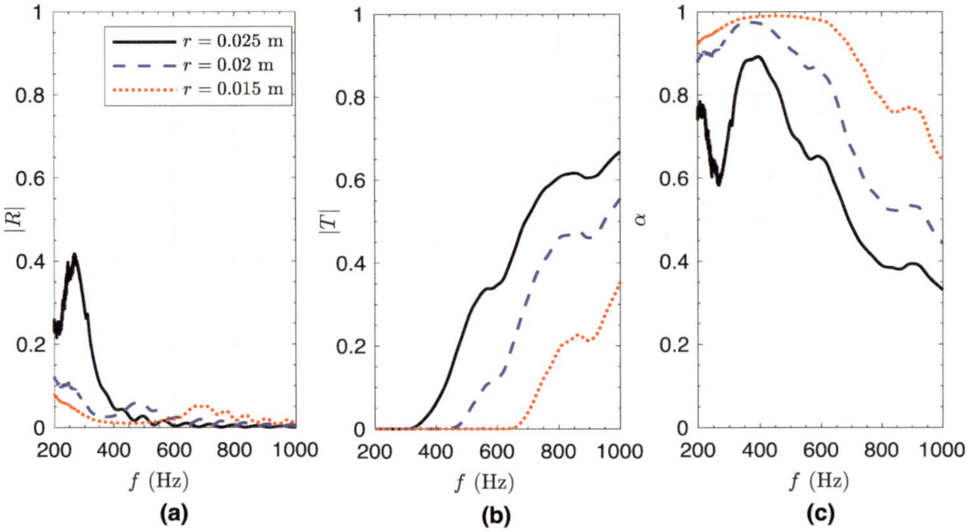

Figure 5.
(color online) amplitude of the (a) reflection coefficients, (b) transmission coefficients and (c) absorption coefficients. Black lines ($r = 0.025$ m), blue dashed lines ($r = 0.02$ m), red dotted lines ($r = 0.015$ m) present the analytical results with a source level around 140 dB (nonlinear case).

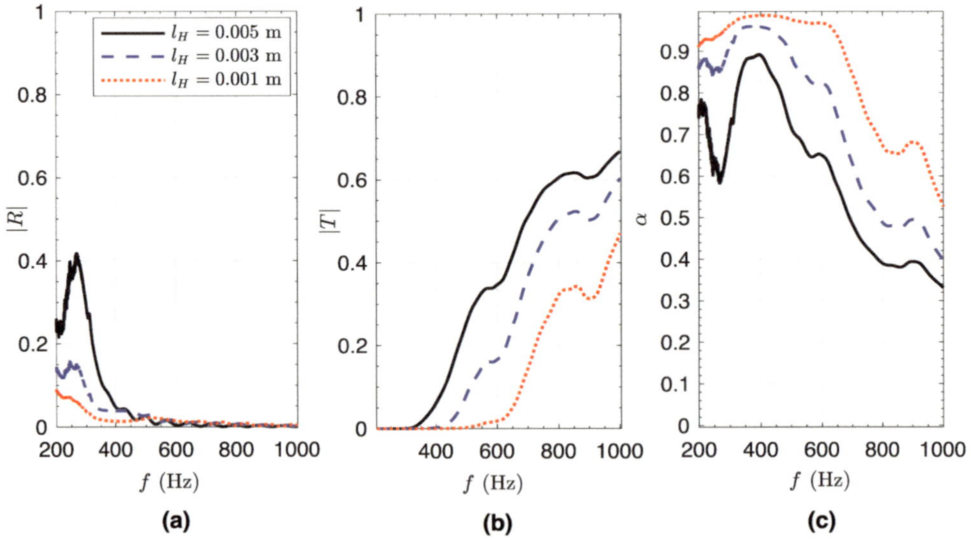

Figure 6.
(color online) amplitude of the (a) reflection coefficients, (b) transmission coefficients and (c) absorption coefficients. Black lines ($l_H = 0.005$ m), blue dashed lines ($l_H = 0.003$ m), red dotted lines ($l_H = 0.001$ m) present the analytical results with a source level around 140 dB (nonlinear case).

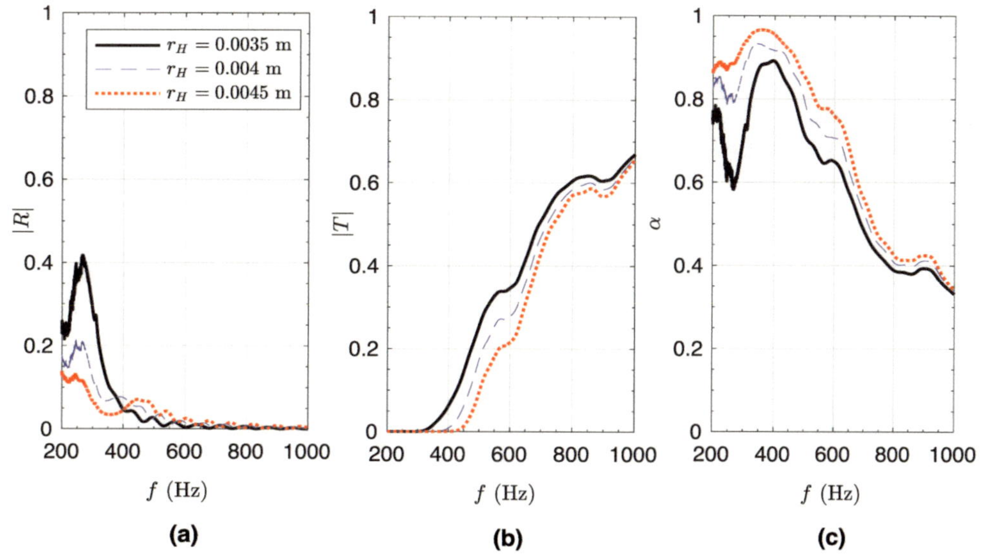

Figure 7.
(color online) amplitude of the (a) reflection coefficients, (b) transmission coefficients and (c) absorption coefficients. Black lines ($r_H = 0.0035$ m), blue dashed lines ($r_H = 0.004$ m), red dotted lines ($r_H = 0.0045$ m) present the analytical results with a source level around 140 dB (nonlinear case).

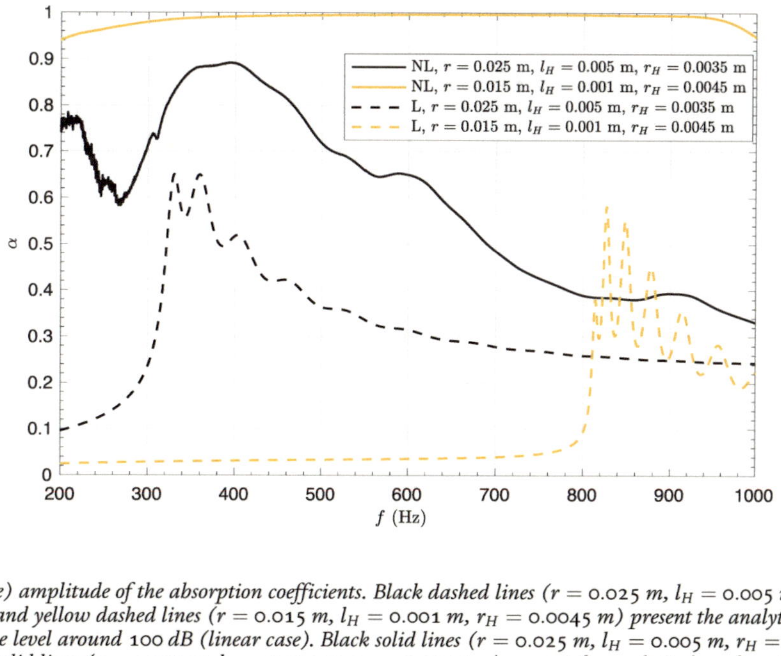

Figure 8.
(color online) amplitude of the absorption coefficients. Black dashed lines ($r = 0.025$ m, $l_H = 0.005$ m, $r_H = 0.0035$ m) and yellow dashed lines ($r = 0.015$ m, $l_H = 0.001$ m, $r_H = 0.0045$ m) present the analytical results with a source level around 100 dB (linear case). Black solid lines ($r = 0.025$ m, $l_H = 0.005$ m, $r_H = 0.0035$ m) and yellow solid lines ($r = 0.015$ m, $l_H = 0.001$ m, $r_H = 0.0045$ m) present the analytical results with a source level around 140 dB (nonlinear case).

optimizing the parameters of the device, we choose $r = 0.015$ m, $l_H = 0.001$ m, $r_H = 0.0045$ m, see **Figure 8**. Now, in a wide frequency range, i.e., 300–900 Hz, the absorption coefficient equal to 1. Even in the range of 200–300 Hz or 900–1000 Hz, the absorption coefficient is not equal to one, however it is greater than 0.9.

5. Conclusions

In conclusion, we have analytically designed a large band amplitude-dependent acoustic absorber. We have studied nonlinear wave propagation in an 1D air-filled waveguide periodically side loaded by holes. We have performed the experiments to highlight the nonlinear losses effect. Transfer matrix theory could analytically capture the experimental phenomenon. Additionally, we analytically tune the parameter of the device and design a large band amplitude-dependent acoustic absorber. Our results concerning nonlinear losses, especially regarding the observed large value of the absorption coefficient in a relatively large bandwidth.

Acknowledgements

I would like to express my most sincere appreciation to my PhD advisors, Vicente Romero-García, Georgios Theocharis, Olivier Richoux, Vassos Achilleos and Dimitri Frantzeskakis. Your guidances are invaluable. Maybe I saved the galaxy in my previous life to meet you as my supervisors. At the beginning of the first year, I did not know anything about research. Even the MATLAB program I could not write well. It is your patience and guidance that make me more and more interested in scientific research. It's like teaching a baby to walk. I am that lucky baby. In the daily life, you also give me a lot of dedicated care and encouragements. Most of my work during the PhD is about theory and simulations, which gives me a good foundation. In the last months of my thesis, you help me to go even further and make my work perfect. Here I would like to thank again Vicente Romero-García and Olivier Richoux for helping me open the door of experiments. I find that the interaction between theoretical studies and experimental verifications is a wonderful process. The authors gratefully acknowledge financial support from the Underwater Acoustics Key Laboratory Stability Support Project (No. J2222006).

Author details

Jiangyi Zhang[1,2]

1 Harbin Engineering University, College of Underwater Acoustic Engineering, Harbin, China

2 le Mans Université, Laboratoire d'Acoustique de l'Université du Mans (LAUM), UMR CNRS 6613, Institut d'Acoustique-Graduate School (IA-GS), CNRS, Le Mans, France

*Address all correspondence to: zhangjiangyi0607@gmail.com

References

[1] Lee D, Nguyen DM, Rho J. Acoustic wave science realized by metamaterials. Nano convergence. 2017;**4**:3

[2] Veselago VG. The electrodynamics of substances with simultaneously negative values of μ and ε. Soviet Physics Uspekhi. 1968;**10**(4):509-514

[3] Liu Z, Zhang X, Mao Y, Zhu YY, Yang Z, Chan CT, et al. Locally resonant sonic materials. Science. 2000;**289**:1734

[4] Liang B, Yuan B, Cheng JC. Acoustic diode: Rectification of acoustic energy flux in one-dimensional systems. Physical Review Letters. 2009;**103**: 104301

[5] Li X-F, Xu N, Liang F, Lu M-H, Cheng H, Chen Y-F. Tunable unidirectional sound propagation through a sonic-crystal based acoustic diode. Physical Review Letters. 2011; **106**:084301

[6] Li J, Fok L, Yin X, Bartal G, Zhang X. Experimental demonstration of an acoustic magnifying hyperlens. Nature Mater. 2009;**8**:931-934

[7] Climente A, Torrent D, Sanchez-Dehesa J. Sound focusing by gradient index sonic lenses. Applied Physics Letters. 2010;**97**:104103

[8] Welter JT, Sathish S, Christensen DE, Brodrick PG, Cherry MR. Focusing of longitudinal ultrasonic waves in air with an aperiodic flat lens. The Journal of the Acoustical Society of America. 2011;**130**: 27892796

[9] Zhao J, Bonello B, Boyko O. Focusing of the lowest-order antisymmetric lamb mode behind a gradient-index acoustic metalens with local resonators. Physical Review B. 2016;**93**:174306

[10] Xu N, He C, Sun X-C, Liu X p, Lu M-H, Feng L, et al. Topologically protected one-way edge mode in networks of acoustic resonators with circulating air flow. New Journal of Physics. 2015;**17**: 053016

[11] Peano V, Brendel C, Schmidt M, Marquardt F. Topological phases of sound and light. Physical Review X. 2015;**5**:031011

[12] Peng Y-G, Qin C-Z, Zhao D-G, Shen Y-X, Xu X-Y, Bao M, et al. Experimental demonstration of anomalous floquet topological insulator for sound. Nature Communications. 2016;**7**:13368

[13] Cummer SA, Schurig D. One path to acoustic cloaking. New Journal of Physics. 2007;**9**:45

[14] Chen H, Chan CT. Acoustic cloaking in three dimensions using acoustic metamaterials. Applied Physics Letters. 2007;**91**:183518

[15] Torrent D, Sanchez-Dehesa J. Acoustic cloaking in two dimensions: A feasible approach. New Journal of Physics. 2008;**10**:063015

[16] Shamonina E, Solymar L. Waves in Metamaterials. New york edition: Oxford University Press; 2009

[17] Henríquez VC, García-Chocano VM, Sánchez-Dehesa J. Viscothermal losses in double-negative acoustic metamaterials. Physical Review Applied. 2017;**8**:014029

[18] Gracía-Salgado R, García-Chocano V, Torrent D, Sánchez-Dehesa J. Negative mass density and density-near-zero quasi-two-dimensional metamaterial: Design and applications. Physical Review B. 2013;**88**:224305

[19] Guild MD, García-Chocano VM, Kan W, Sánchez-Dehesa J. Acoustic metamaterial absorbers based on multilayered sonic crystals. Journal of Applied Physics. 2015;**117**:114902

[20] Romero-Garcia V, Theocharis G, Richoux O, Merkel A, Tournat V, Pagneux V. Perfect and broadband acoustic absorption by critically coupled sub-wavelength resonators. Scientific Reports. 2016;**6**:19519

[21] Theocharis G, Romero-Garcá V, Richoux O, Tournat V. Limits of slow sound propagation and transparency in lossy, locally resonant periodic structures. New Journal of Physics. 2014; **16**:093017

[22] Ma G, Yang M, Xiao S, Yang Z, Sheng P. Acoustic metasurface with hybrid resonances. Nature Materials. 2014;**13**:873-878

[23] Lapine M, Shadrivov IV, Kivshar YS. Colloquium: Nonlinear metamaterials. Reviews of Modern Physics. 2014;**86**:1093

[24] Shalaev MI, Myslivets SA, Slabko VV, Popov AK. Negative group velocity and three-wave mixing in dielectric crystals. Optics Letters. 2011; **36**(19):38613

[25] Khurgin JB. Optical parametric oscillator: Mirrorless magic. Nature Photonics. 2007;**1**:446447

[26] Popov AK, Slabko VV, Shalaev VM. Second harmonic generation in left-handed metamaterials. Laser Physics Letters. 2006;**3**(6):293

[27] Hamilton M, Blackstock DT. Nonlinear Acoustics. San Diego, CA: Academic Press; 1998

[28] Sánchez-Morcillo VJ, Pérez-Arjona I, Romero-Garcá V, Tournat V, Gusev VE. Second-harmonic generation for dispersive elastic waves in a discrete granular chain. Physical Review E. 2013; **88**:043203

[29] Jiménez N, Mehrem A, Picó R, García-Ra LM, Sánchez-Morcillo VJ. Nonlinear propagation and control of acoustic waves in phononic superlattices. Comptes Rendus Physique. 2016;**17**: 543554

[30] Zhang J, Romero-García V, Theocharis G, Richoux O, Achilleos V, Frantzeskakis DJ. Second-harmonic generation in membrane-type nonlinear acoustic metamaterials. Crystals. 2016; **6**(8):86

[31] Boechler N, Theocharis G, Daraio C. Bifurcation-based acoustic switching and rectification. Nature Mater. 2011;**10**:665

[32] Devaux T, Tournat V, Richoux O, Pagneux V. Asymmetric acoustic propagation of wave packets via the self-demodulation effect. Physical Review Letters. 2015;**115**:234301

[33] Khajehtourian R, Hussein MI. Dispersion characteristics of a nonlinear elastic metamaterial. AIP Advances. 2014;**04**:124308

[34] Donahue CM, Anzel PWJ, Bonanomi L, Keller TA, Daraio C. Experimental realization of a nonlinear acoustic lens with a tunable focus. Applied Physics Letters. 2014;**104**: 014103

[35] Manktelow K, Leamy MJ, Ruzzene M. Multiple scales analysis of wavewave interactions in a cubically nonlinear monoatomic chain. Nonlinear Dynamics. 2011;**63**:193

[36] Boechler N, Job S, Kevrekidis PG, Theocharis G, Porter MA, Daraio C. Discrete breathers in one-dimensional

diatomic granular crystals. Physical Review Letters. 2010;**104**:244302

[37] Averkiou MA, Lee YS, Hamilton MF. Self-demodulation of amplitude and frequency modulated pulses in a thermoviscous fluid. The Journal of the Acoustical Society of America. 1993; **94**(5):28762883

[38] Vos HJ, Goertz DE, de Jong N. Self-demodulation of high-frequency ultrasound. The Journal of the Acoustical Society of America. 2010;**127**(3): 1208-1217

[39] Sugimoto N, Masuda M, Ohno J, Motoi D. Experimental demonstration of generation and propagation of acoustic solitary waves in an air-filled tube. Physical Review Letters. 1999;**83**:4053

[40] Achilleos V, Richoux O, Theocharis G, Frantzeskakis DJ. Acoustic solitons in waveguides with helmholtz resonators: Transmission line approach. Physical Review E. 2015;**91**: 023204

[41] Zhang J, Romero-García V, Theocharis G, Richoux O, Achilleos V, Frantzeskakis DJ. Bright and gap solitons in membrane-type acoustic metamaterials. Physical Review E. 2017; **96**:022214

[42] Atig M, Dalmont J, Gilbert J. Termination impedance of open-ended cylindrical tubes at high sound pressure level. Comptes Rendus Mécanique. 2004;**332**(4):299-304

[43] Buick JM, Skulina AMD, Campbell D, Dalmont JP, Gilbert J. Investigation of non-linear acoustic losses at the open end of a tube. The Journal of the Acoustical Society of America. 2011; **129**(3):1261-1272

[44] Zhang J, Romero-García V, Theocharis G, Richoux O, Achilleos V, Frantzeskakis DJ. High-amplitude sound propagation in acoustic transmission-line metamaterial. Applied Physics Letters (Editors' Pick). 2021;**118**:104102

[45] Dalmont JP. Study and Realization of Acoustic Impedance Sensors. Application to Measurements of Lumped Elements. Study and Realization of an Anechoic Termination for Low Frequencies, Ph.D. thesis. University of Maine, HAL; 1988

[46] Song BH, Bolton JS. A transfer-matrix approach for estimating the characteristic impedance and wave numbers of limp and rigid porous materials. The Journal of the Acoustical Society of America. 2000;**107**:1131

[47] Pierce AD. Acoustics: An Introduction to its Physical Principles and Applications. New York: McGraw-Hill; 1981

[48] Zwikker C, Kosten CW. Sound Absorbing Materials. New York: Elsevier; 1949

[49] Temiz MA, Tournadre J, Arteaga IL, Hirschberg A. Non-linear acoustic transfer impedance of micro-perforated plates with circular orifices. Journal of Sound and Vibration. 2016;**366**:418-428

[50] Disselhorst J, Wijngaarden LV. Flow in the exit of open pipes during acoustic resonance. Journal of Fluid Mechanics. 1980;**99**(2):293-319

[51] Peters M, Hirschberg A. Acoustically induced periodic vortex shedding at sharp edged open channel ends: Simple vortex models. Journal of Sound and Vibration. 1993;**161**(2):281-299

[52] Kalozoumis PA, Richoux O, Diakonos FK, Theocharis G. Invariant

currents in lossy acoustic waveguides with complete local symmetry. Physical Review B. 2015;**92**:014303

[53] Dubos V, Kergomard J, Keefe D, Dalmont J-P, Khettabi A, Nederveen K. Theory of sound propagation in a duct with a branched tube using modal decomposition. Acta Acustica united with Acustica. 1999;**85**(2):153-169

[54] Dalmont J-P, Nederveen CJ, Joly N. Radiation impedance of tubes with different flanges: Numerical and experimental investigations. Journal of Sound and Vibration. 2001;**244**(3): 505-534